<u>JOE</u> !

WHAT A CAREER YOU'VE
HAD ! YOU'VE HAD
A BIG PART IN CHANGING
THE WORLD. IT'S BEEN
MY PLEASURE TO KNOW
YOU !

JOHN

To:
Pam, Erin, Al, Amber, Patrick, James, Parker,
Liam, and Ginger

Portions of this book were previously published by Daniel Nenni in his excellent semiconductor BLOG SemiWiki – All Things Semiconductor. SEMIWIKI.com

This second edition was published in the United States by Qiworks Press. Printed in the United States by Ingram Spark. Cover design and illustrations by Al Perry.

ISBN 9780578831343

Table of Contents

Silicon Valley

The way I saw it

John East

Qiworks
Press

Preface

As I was editing this book, I happened to read a press release about the latest Apple chip. Apple had just announced that the integrated circuit that powers their latest Mac – the M1 chip - contained 16 billion transistors -the equivalent of about four billion gates. * I'm sure that there were a few million Americans who read that and said, "What's a gate?" Another few million said, "Who cares?"

The earliest serious work on semiconductor technology started in the 1930s. The concept of an integrated circuit originated, thanks to Jack Kilby and Bob Noyce, in 1958 and 1959. The first functional integrated circuit that was worthy of the name came in 1961. That first circuit didn't contain a billion gates – or a million gates - or even a hundred gates. It contained four gates. It took two years from the time Robert Noyce came up with the idea of what could be done and how to do it until the time the job had actually been accomplished.

---

* A "gate" is a small set of transistors that can work together and "think" at least a little. With a handful of gates hooked together, for example, you can build an adder – a unit that can add two numbers.

It would be nice to say that after the first IC had been made, the technology started to leap forward at an astonishing pace. Wrong. It didn't. Progress was slow and hard to come by for the next couple of decades. The concepts were in place, but a huge number of obstacles remained to be overcome.

It would be even nicer to say that I was the guy who made it happen – who addressed those obstacles one by one clearing a path for the Apples of the world to crank out their four billion gate marvels. Wrong again. I wasn't. But I can say that I was hanging around with the people who made it happen, and that looking back, it was a fantastic voyage.

This book is not intended to be a history of Silicon Valley. That job would be far above my pay grade. It's simply a collection of stories --- stories about events I look back on that now give me the urge to tell someone else about them. To make the stories more readable, I've included short bios of the cast of characters involved in the stories. Without those bios, it might be hard for readers to have the proper perspective.

Note: The terms "semiconductors", "integrated circuits", "ICs", "circuits" and "chips" are used more or less interchangeably in this book. They all mean roughly the same thing.

Context

When I was a grad student at the University of California at Berkeley, I worked in the computer room in Barrows Hall. Our computer at that time was an IBM 1460. How powerful was an IBM 1460? It had 8K bytes of memory (It could remember eight thousand minuscule pieces of information). Today's computers have on the order of a million times more. To put things into perspective, an average length song that you might have on your iPod or MP3 player needs 4M bytes or so (four million minuscule pieces of information) – 500 times more than the 1460 had. That means that the 1460 – which I viewed as a very, very powerful computer in those days -- could store just a couple of notes of your favorite song. Or maybe you prefer thinking in terms of photographs? You know that picture of your nephew? The one where he has the big, toothy grin? The IBM 1460 had enough storage to hold about one tooth of that picture.

Was that problem really so hard to solve? Why not just put together a lot of IBM 1460s? That would still do the job, right? Wrong. Not really. There would be a few snags. The entire 1460 system took up a space about the size of a small bedroom. Putting a million or so of those together? No way!! But – the integrated circuit had recently been invented.

What about making a really, really big IC chip? Wouldn't that be better? Nice try. That chip would be approaching the size of a football field.

That's bad!! The technology doesn't exist even today to make a chip that's much bigger than a square inch or so -- forget one that's the size of a football field. But – those problems pale compared to the big problem -- power. A typical gate back in the day dissipated about 50 milliwatts. The high-speed gates were more than 100 milliwatts. Four billion of them? That's about 500 megawatts!!! What does all that mean in plain English? Do you remember the Chernobyl nuclear reactor melt-down disaster in 1986? That nuclear power plant was rated at 1000 megawatts.

Let's say you were using this football-field-sized chip in a new phone. When you went out to buy that phone, you'd have to buy a Chernobyl-sized nuclear reactor to power it. And – since even today battery technology doesn't come anywhere near to being able to provide that much power - you'd have to carry that nuclear reactor with you wherever you went!! Luckily the reactor could power two phones so you could put yourself on some sort of family plan. You could get a phone for your teenaged daughter as well.

Oh. One other thing. There's the little matter of price. The first available IC gates had a list price of $120 each. So -- the four billion gate chip in your new phone would cost five hundred billion dollars. So much for your family plan.

How did we get from the IBM 1460 to where we are today? Step one: the integrated circuit!! The IC!! The chip!! Step two: lots and lots of progress! The 1460 didn't use integrated circuits -- they hadn't been invented yet when the 1460 was first designed. And when the first ICs were made, they weren't any better in most regards than the discrete circuits that they eventually replaced.

The invention of the IC was the beginning, but huge amounts of progress were required to get us to where we are today. Today's ICs are conceptually made the same way that Bob Noyce envisioned with one very significant exception --- today's circuits are incredibly more complex!! It's commonplace today to develop circuits that are a billion times more complex than Noyce's original IC. Now that's progress!!!!

Life without the integrated circuit? Hard to imagine. No cellphone. No personal computer. No internet. No email. No Facebook or Twitter or Snapchat. No instant messages. No Google. No Netflix. Sure --- You could watch TV. But when you wanted to change the channel you'd have to get up off the couch, walk over to the TV, and turn the knob yourself. Maybe that wouldn't be a big problem, though. You'd probably only have the choice of four or five channels. The huge companies that drive America's economy today owe their existence to the modern IC. Apple – certainly!! Google, Facebook, and Microsoft? Of course. Even Amazon wouldn't exist today without the IC. That's worse than it sounds. When I was young; steel, automobiles and oil drove the American economy. Not anymore. Those industries are all floundering. Today high-tech drives our economy. No high tech would mean no jobs. No tax base to pay for our five trillion dollar federal budget. Not a pretty picture!!!!

We needed that progress!!

Warning -- all the numbers and examples above are very, very approximate.

I made every effort to write this book so that it could be easily read by non-engineers. Sometimes I succeeded. Sometimes I failed. The good news is that the episodes are more or less independent. That means that if you find yourself reading an episode that you're not understanding or not enjoying (Or both), just skip it and go on to the next one. Very little will be lost. But --- be sure to read episode #20

Episode 1

Part 1

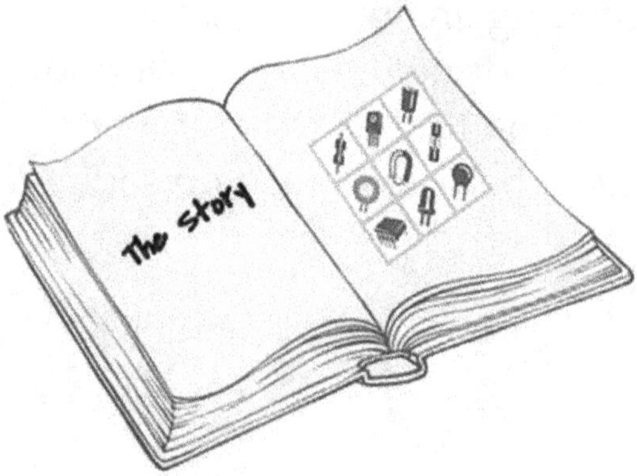
The story

The Story

Day 1

In 1967 I was a grad student at Cal Berkeley. In December of that year my wife to be and I got engaged to be married. I was supposed to get my master's degree in December of '68, but once we worked out all the details, we realized that I'd have to go to school over the summer of '68, get the degree in September, and get a job. We were broke and couldn't afford the extra three months of expenses with little or no income. Berkeley was set up with biannual college recruiting programs during which corporations would come in to interview prospective new hires. One of the sessions was in March and one was in November. My original plan was to go through the college recruiting process in the November session, but the wedding plans changed that.

Since I wouldn't be ready to go to work until September, the March recruiting session seemed too early. So --- how to get a job? That was the question.

I wrote 40 or 50 letters. There was a college placement handbook that had the addresses of the important companies. I wrote to them basically saying "Dear Sir, you don't know me, but I want a job." I got back just three responses which was a little depressing. One was from IBM, where I then interviewed and didn't get a job offer. One was from HP where I interviewed and didn't get a job offer. But one was from Fairchild. All I knew about them --- or thought I knew --- was they made cameras. (The official company name at that time was Fairchild Camera.) I interviewed with them and they were excited about me. They brought me back a short while later to have lunch with two of their executives: Jerry Briggs – an HR guy (Except what is now called Human Relations was called Personnel in those days) -- and Gene Flath - a product line manager. That was my first business lunch. It turned out that in those days, business lunches involved large quantities of martinis and the like. They thought I was the greatest guy in the world (Possibly because of the martinis?) and they offered me a job on the spot. This was in roughly May of '68. They knew that I wasn't going to be done with school until September, but they said, "That's not a problem. We'll wait for you. You're going to be wonderful. In fact, you don't even need to communicate with us in the interim. The day before you're done, just call us and we'll make arrangements for you and everything will be great." Then they both gave me their business cards.

When I had one day to go ---that is I had just taken my last final and was ready to go to work ---- I picked up the phone and called Fairchild HR. A lady answered the phone. I asked, "Can I please speak to Jerry Briggs?" The lady who answered the phone said, "You must be mistaken. Are you sure you called the right number? There's no Jerry Briggs here and I've never even known a Jerry Briggs." We debated for a while and after a bit I asked her "Well, how long have you been there?" It had been a couple of months. I asked to speak to someone who had been there in May. There was no one. The department had turned totally over between the time of the offer in May and my call in September. I thought to myself, "Wow. Are you kidding? But that's not a problem because I've got Gene Flath's card as well. I'll just call Gene Flath." So --- I called Gene Flath's number and got a secretary. She said, "You must be mistaken. Are you sure you called the right number? There's no Gene Flath here and there's never been a Gene Flath here in all of the time I've been here." "Well, how long have you been there?" "A couple of months." Uh-Oh!! Red flag!! I asked myself, "What the heck is going on here?" I needed that job! Fortunately, I had the offer letter. I called the HR department again and told them so. Some guy who I had never met said "Well, okay, we'll honor it. Come in at 9:00 on Monday morning and we'll figure out what to do with you."

What the heck was going on? I found out later that Bob Noyce, the President of Fairchild, had just left to form Intel Corporation and taken a cadre of the really good people with him. Sherman Fairchild (the Chairman of the Fairchild Board of Directors) had brought in Les Hogan from Motorola to be the new CEO. Hogan, then, brought in eight of his top lieutenants to help him run things. They were referred to as 'Hogan's Heroes'. (That was the name of a popular TV show in those days.) Hogan and his Heroes proceeded to fire about a third of the upper ranks. Roughly another third of the upper ranks said to themselves, "Well, wait a minute. If I stay around, they're going to fire me, too." So, they left as well. Everything had turned over in that four-month window. Gene Flath had gone to Intel. I have no idea what happened to Jerry Briggs. When I got there nobody knew what was going on. Nobody knew who their boss was. What a zoo it was, but that made it almost seem like fun. One thing that was particularly notice-able was ---- in the other companies where I had interviewed the managers were 40-year- old or 50-year-old people. Today that doesn't seem very old, does it? But then it seemed ancient. "You mean, I've got to be around for twenty years be-fore I can get a manager job? That's terrible." At Fairchild, the managers were kids. They were 25 and 30 years old. And not only were they kids, but they were also kids viewed as being experts in their field because the field was that young.

I thought, "I'm going to like this place."

Episode 1

Part 2

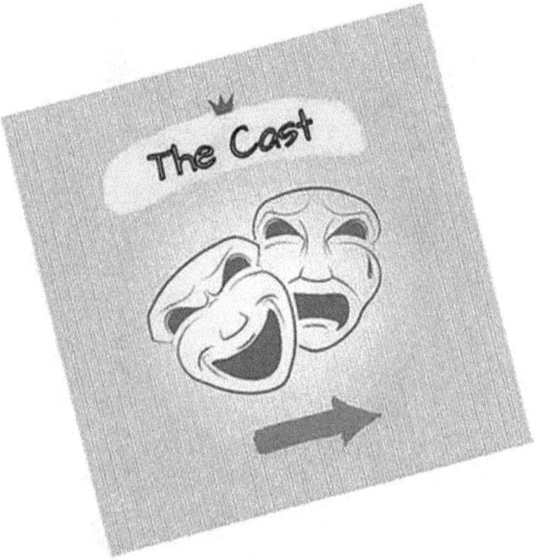
The Cast

TheCast

Silicon Valley

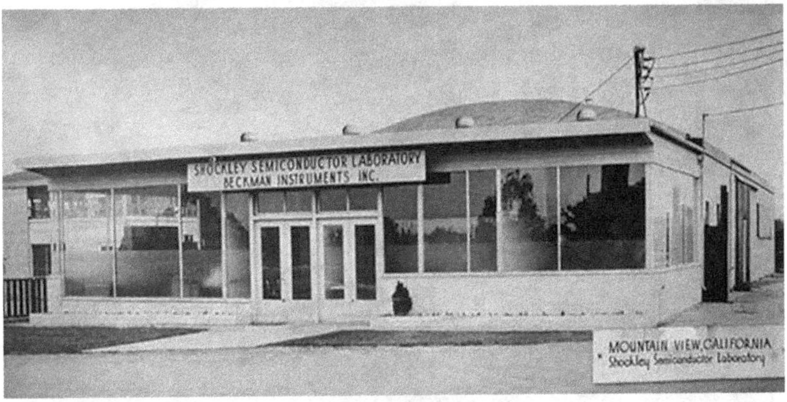
SHOCKLEY SEMICONDUCTOR LABORATORY
BECKMAN INSTRUMENTS INC.
MOUNTAIN VIEW, CALIFORNIA
Shockley Semiconductor Laboratory

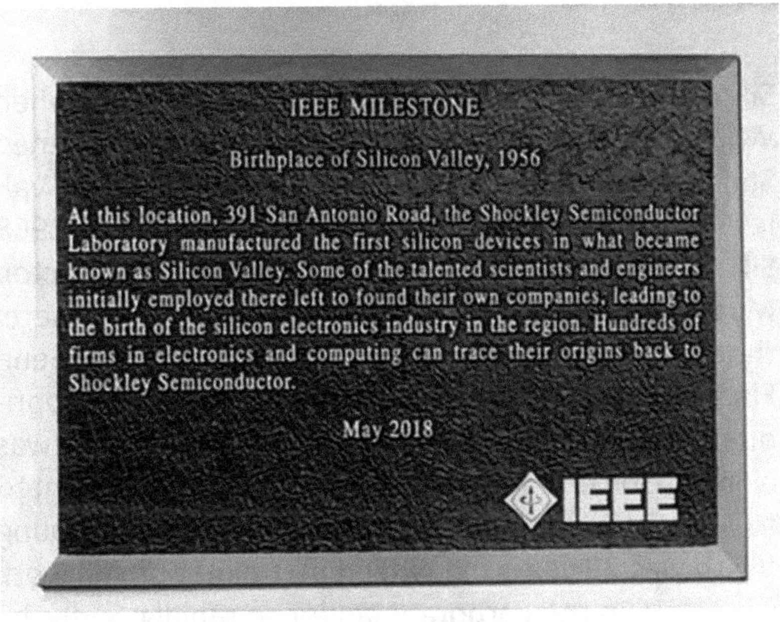
IEEE MILESTONE
Birthplace of Silicon Valley, 1956
At this location, 391 San Antonio Road, the Shockley Semiconductor Laboratory manufactured the first silicon devices in what became known as Silicon Valley. Some of the talented scientists and engineers initially employed there left to found their own companies, leading to the birth of the silicon electronics industry in the region. Hundreds of firms in electronics and computing can trace their origins back to Shockley Semiconductor.
May 2018
IEEE

Images 1 & 2

Fairchild Semiconductor

Images 3

Silicon Valley was born (although not yet named) when William Shockley moved to California and opened Shockley Semiconductor. By the time I came to the Valley in 1968 Shockley Semiconductor was gone. In 1968 Silicon Valley was all about Fairchild Semiconductor. My first job was at Fairchild. When I interviewed there, the first person I met was a manager named Gene Flath. Gene was the product manager of the Proprietary Integrated Circuits department. To me that was akin to God. He impressed me a lot. I really wanted to work at a company where you could be so young (Flath was 29 years old when I met him) but still work your way up the corporate ladder so rapidly. Sadly, by the time I arrived at Fairchild, Flath was gone. He was one of the very first hires made by Intel following their founding in 1968. Even though we've both been in Silicon Valley for the last 52 years, I've never run into him again. Gene -- if you happen to read this, give me a call. Lunch is on me!!

The first of the three pictures is of the original (and only) Shockley Semiconductor building. It was located at roughly the corner of El Camino and San Antonio Road in Mountain View. That was the site of the first silicon in Silicon Valley. It was torn down a few years ago. There's a Hyatt Hotel there now. The hotel restaurant is called "the Fairchild Public House".

The second picture is of a historical landmark plaque which sits today at the site of the Shockley building. The plaque credits the site as being the origin of Silicon Valley. In my mind this is right. Some give that credit to the famous HP Garage in Palo Alto, but in my mind that title – the origin of Silicon Valley – should go to a *silicon* company. HP was (and still is) a great company, but it was never truly a silicon company. Shockley was all about silicon.

The third picture is of the original Fairchild building at roughly the corner of San Antonio Road and Charleston Road. 844 E. Charleston Road, Palo Alto. It's just a mile or so from the Shockley building. The Fairchild building is still there. It's where the integrated circuit was invented.

Every now and then I drive by, stop, and say thank you.

John and Pam East

Images 4 and 4a

Two months before my first day on the job at Fairchild, Pam Mattson and I got married. We were both going to school at Cal-Berkeley at the time. I was a bit of a cradle robber -- I was finishing my master's degree but Pam had just finished her sophomore year. We were dead broke -- I had to borrow money in order to get a haircut before my first day at Fairchild. But -- at least we were secure in the fact that I had a good job lined up the moment I finished the degree. -- Or so we thought --- One day before I was supposed to start, the job disappeared! I had never met anyone at Fairchild who still worked there. I knew nothing about integrated circuits. I had no idea what "wafer-sort" was. I thought "class" was a place you went when you wanted to learn something. (It wasn't) I should have been scared to death. Somehow, I wasn't. It's great to be young!

Les Hogan

Images 5

Les Hogan was the CEO of Fairchild when I got there in 1968. He had done post-graduate studies at Lehigh University where he obtained a Ph.D. in Physics. He joined Bell Labs in 1950 where he worked under Bill Shockley, the inventor of the transistor. More about Shockley in episode 3. In early 1968 he was the vice president and general manager of the semiconductor operation of Motorola -- a large manufacturer of semiconductor products. In the summer of 1968 he moved to Fairchild Camera & Instrument as Chairman and CEO, taking eight senior executives (nicknamed *Hogan's Heroes* after the popular TV show of that name) with him. This move caused Motorola to sue Fairchild (unsuccessfully) for theft of trade secrets. More importantly to me, though -- it also caused the sheer bedlam that was running amok inside of Fairchild when I got there. Hogan joining Fairchild was at the roots of the *"Off with Their Heads"* episode.

I barely interacted with Les. He was the CEO. I was the lowest level engineer. Still, the few times I had the opportunity to be around him, he was approachable and friendly. I liked Les. I felt badly when he got the axe in 1974 although when you read the episode, *"Layoffs Ala Fairchild,"* you may well conclude that he had it coming.

11

Episode 2

Part 1

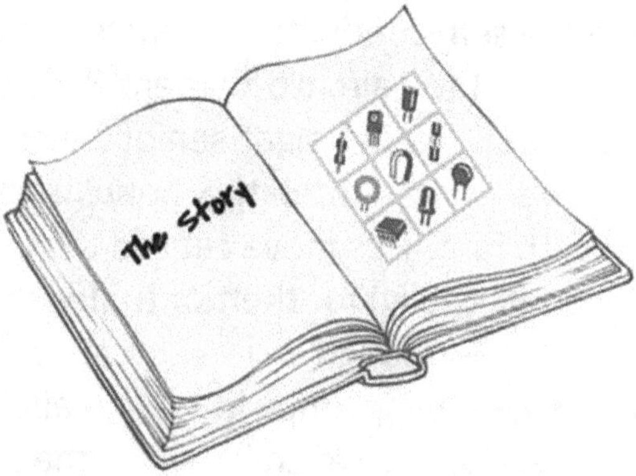
The story

The Story

Off with Their Heads

I started out as a supervisor in the wafer sort and class area. Today you'd call those probe and final test. My first boss, a man named Les Faerber who I had never met, greeted me in the lobby, got a smock for me, took me into the test area and introduced me to the ladies (All the operators were women in those days). I had no idea what they were doing. I didn't even know that integrated circuits were made on wafers much less that those wafers needed to be "sorted." Then he said, "I've got a meeting. Gotta go." And he left. Terrifying!!! I was standing there trying to act as though I knew what was going on (I didn't have a clue) when a really aggressive guy with a British accent came charging in. "Who's the supervisor here?!!!" "Why isn't the waterfall running?!!!" And of course, I thought to myself, "Who is this guy? What's a waterfall? What am I doing here?!!!" It was John Carey. He was the operations manager for all integrated circuits. He wasn't a patient man! He scared me to death the first few times I dealt with him, but after a while I grew fond of him. That was a shame, because very soon he would fall victim to *"Off with their heads."*

That first afternoon a technician named Jack Drury was trying to teach me a little of what we were

13

doing in wafer sort and class. I'm sure it crossed his mind to wonder why an experienced technician like him was now working for an acne-faced college kid who had no clue about anything that mattered, but he tried to be helpful. We stood in front of a prober and watched a wafer being tested (sorted). I was impressed with the wafers. "Wow --- you can make hundreds of these IC things at a time. That's cool!" Each wafer had a few hundred "dice" on it. Our job was to test them and identify the good ones. (At Fairchild, we called the individual ICs "dice." A single IC chip was a "die." At most other companies they called them "chips.") The prober would stop on each die. The tester would flash cool looking lights on and off for a few seconds, and then a small mechanical arm would put a little red dot on the die. It seemed pretty efficient. I asked Jack what the little red dot was for. He said, "Oh. That's the inker. We put a red ink dot on each die that doesn't work right."

Me: "Oh. I see. That's cool. But – every one of the dice has a little red dot."

Jack: "So?"

Me: "So all of these wafers that were already sorted have red dots on all their dice as well."

Jack: "What's your point?"

Me: "Duh"

Jack: "That's a TTL lot. The lot is zeroing out. Big deal. That happens all the time to TTL."

I think it was a 50-wafer lot. Each wafer probably had 500 dice on it. So --- the first 25,000 integrated circuits that I ever saw were all thrown out. Nobody seemed to care.

Maybe that's what led up to *"Off with their heads?"*

What was *"Off with their heads?"* Well, as I described in "Day One," many heads had already rolled after Hogan's Heroes arrived but before I got there. I heard some people saying good things about a man named Tom Bay. I asked what job he was in. "Oh. He's gone. Fired a month ago." Tom was formerly the VP of marketing and, as far as I knew, the first victim of *"Off with their heads!"* Then, shortly after I got there, the VP of sales, Jerry Sanders, was fired. Jerry, of course, then founded AMD and went on to a great career. Then, John Carey (The guy who wanted the waterfall running) was fired. He went with Jerry Sanders to found AMD and later became the CEO of IDT. Carey was replaced by a man named John Husher, but he was in and out of there so fast that I never got to meet him. Bob Noyce, Gordon Moore, and Andy Grove had already left to form Intel.

Charlie Sporck gone too. Charlie had left to be CEO of National Semiconductor. Charlie took Floyd Kwamme, Don Valentine, and Pierre Lamond (all of eventual venture capital fame) with him. Venture capital? -- Gene Kleiner was gone as well. Gene was one of the Traitorous Eight – the group of eight men who defected from Shockley Semiconductor in order to form Fairchild - but went on to be the head of what is probably the most famous Venture Capital firm of all - Kleiner Perkins. In fact, every one of the Traitorous Eight founders left under various circumstances. (More about the Traitorous Eight in Episodes 3 and 10) There were people disappearing left and right. Sometimes you didn't know if they were fired or if they just quit. One morning you'd come in and they were gone. Why? Where? How? Who knew? Who was doing all this firing? Who knew? It seemed as though a Vice President and Hogan's Hero named Gene Blanchette was at the root of a lot of it, but before long, Blanchette himself disappeared. I don't remember how or why. I probably never knew. And then, in 1974 the coup de grace. Les Hogan - the CEO – the author of off with their heads - was gone too. Fired. There was a popular song in 1967. "White Rabbit" by The Jefferson Airplane. Its lyrics put a new spin on Alice in Wonderland. -- "And the red queen (said) *'Off with their heads!'"* --- Wow! An apt description of what I'd just walked into.

Have I rediscovered Alice in Wonderland? Did I go down the rabbit hole? And where did I put that hookah?

Episode 2

Part 2

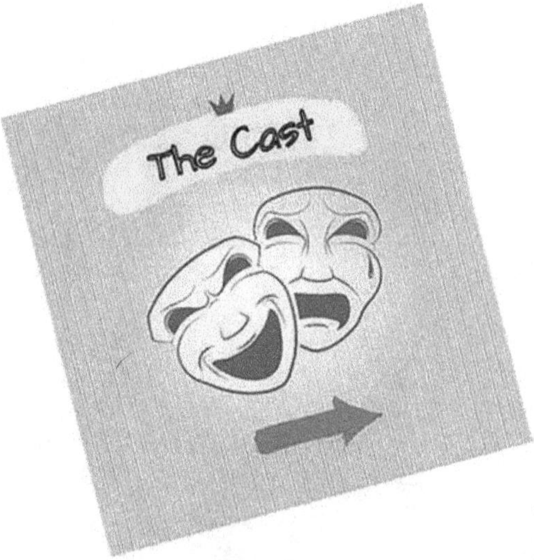
The Cast

The Cast

Charlie Sporck

Image 6

Charlie Sporck studied mechanical engineering at Cornell University, graduating with a bachelor's degree in 1950. He then took a job working at General Electric in New York. In 1959 he accepted a job as a production manager at Fairchild. Charlie had a story similar to mine in episode Day One. He was excited about his new job. He quit his job at GE and moved his family from New York to Mountain View. Then he showed up at Fairchild for his first day at work. No one knew who he was. No one knew that he had been hired. He attributed that to the fact that there had been some heavy drinking going on by the people who interviewed him and gave him the job offer. Charlie overcame that and rapidly worked his way up to the top job in manufacturing at Fairchild reporting to Bob Noyce. (You'll soon read much more about Bob Noyce.)

In 1967 he was recruited by National Semiconductor to be their general manager. When he left Fairchild, he took Floyd Kwamme, Fred Bialek, and Pierre Lamond to National with him. They were all very important to Fairchild's efforts at the time. That created a vacuum at Fairchild that was never really filled. Charlie was a physically large man - probably 6'5" - with a demeanor that could be very intimidating! You didn't dare mess with Charlie Sporck!! His book, *"Spinoff,"* is a fun read – full of stories about how it was in the early days.

John Carey

Image 7

John Carey was born in England and studied at the University of Liverpool ending with a degree in electrical engineering. He later moved to California to join Fairchild Semiconductor. John was the Operations Manager of Integrated Circuits when I first joined Fairchild in 1968. He was intimidating!! I was scared to death of John. He was my first exposure to the management by walking around type of manager. At random times during the day John would come barging into the test area. No - not **sneaking** in. **Barging** in!!! Then he would find something wrong. That's usually not too hard to do, but John was extraordinarily good at it! And once he found it, he made sure that you knew it had better get fixed in a hurry and never happen again. In 1968, having been fired from Fairchild -- a casualty of the new "Hogan's Heroes" regime -- he and seven others formed AMD. There John worked for Jerry Sanders and ran all of operations. At AMD the people who worked for him were smart. They determined that John had a certain routine - he always visited the various fabs and test areas in the same order. So – as soon as John left an area, the hapless victim would know where John was headed. He would call the victim-to-be and warn him, "He's coming!!!" But – it didn't always help. From AMD, John went on to become the CEO of IDT. I grew to love John Carey. He was a unique human being.

Traitorous Eight

Image 8

The Traitorous Eight were the eight men who founded Fairchild Semiconductor. They all worked together
at Shockley Semiconductor. In episode #3 you'll learn
more about William Shockley. He was a difficult man
to work for! Eventually these men tired of working
for Shockley and resigned. Gene Kleiner's father was
involved in investment banking. He introduced the
group to Arthur Rock, another banker, who then introduced them to Sherman Fairchild. Sherman Fairchild
had founded Fairchild Camera and was chairman of its
board of directors. The Traitorous Eight struck a deal
with Sherman Fairchild. Fairchild agreed to finance
them as their own independent company so long as

he had an option to buy them out if they did well. They did well. He bought them out.

Gene Kleiner is third from the left in this picture. Gordon Moore, of Moore's Law fame, is at the far left. Robert Noyce, the inventor of the integrated circuit, is front and center. Jay Last, who actually **made** the first integrated circuit, is at the far right. The others, from the left, are Sheldon Roberts, Victor Grinich, Julius Blank, and Jean Hoerni. I once delivered a podcast dealing with the Traitorous Eight. (You can find it by Googling "Podcasts Archives SemiWiki") It's a fun story, but too long to tell here.

Episode 3

Part 1

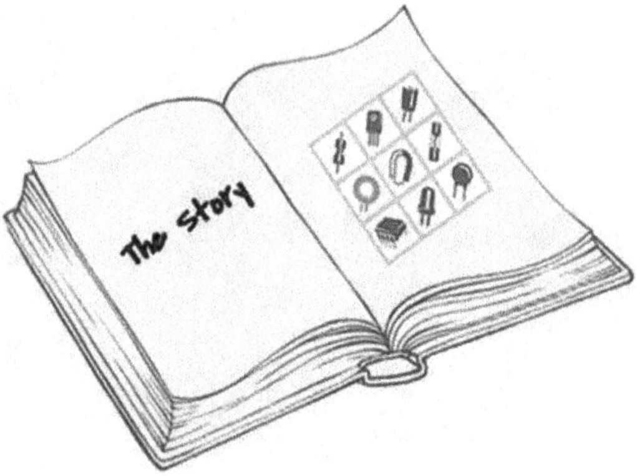
The story

The Story

The Genius Sperm Bank

How did it happen? How did Fairchild transform over a decade into the "off with their heads" culture? To understand that, you need to know a little about the William Shockley story. William Shockley was born in London in 1910. He moved to Silicon Valley when he was three. Of course, it wasn't called Silicon Valley then. There was no silicon in Silicon Valley until he brought it here in 1955. He was a problem child from the get-go. He had a terrible temper -- generally driving his parents nuts. But --- He was smart!!! He could count to four before he turned one year old. In fact, those who knew him later in life generally agreed that he was the smartest person they'd ever met. He studied at Cal Tech and then did his PhD at MIT. Then he took a job at Bell Telephone Laboratories in New Jersey.

Shockley built such a reputation for brilliance in his early years at Bell that the government "borrowed" him during World War II to assist in various planning and strategic efforts. At one time he held the title of Director of Research for the US war efforts. At the end of the war, he was awarded the National Medal of Merit for contributions made to our war effort even though he had remained, in fact, a civilian throughout the war.

After the war, Shockley returned to Bell Labs where he worked on semiconductor technology with Walter Brattain and John Bardeen. The concept of a "transistor" (It hadn't been named that yet) had existed for several years in people's minds, but no one had succeeded in building one. In 1947, Brattain, Bardeen and Shockley succeeded. They built the first functional transistor. Most of the work, in fact, was done by Bardeen and Brattain, but Shockley was happy to accept most of the credit. To be fair, Shockley did have the best understanding of what they had just accomplished and subsequently was generally recognized as being the foremost expert in transistor physics of the times. In 1956, the three received a Nobel Prize for their discovery.

In 1955 Shockley decided that he didn't want to work for a salary any longer. He reasoned that, with his superior intellect and the head start he had at understanding semiconductors, he should be able to make a fortune by starting his own semiconductor company. Accordingly, he moved to Palo Alto, his old home, and opened Shockley Semiconductor. He hired the best and the brightest engineers and scientists that he could find (Most of them went on to have fabulous careers) and set out to conquer the semiconductor world. There was just one problem --- in a management role he was an overbearing tyrant. He drove all his employees nuts. Eventually they bailed out. In particular, a group of eight brilliant scientists and engineers led by Robert Noyce and including Gordon Moore gave notice and left en masse. Shockley dubbed them the Traitorous Eight. Because of the mass exodus coupled with a major strategic error (Focusing on 4-layer diodes instead of transistors) Shockley Semiconductor eventually went under.

I met Bill Shockley at Stanford at some time during the 70s. He was giving a lecture at Stanford that was open to the public. I wish I could remember exactly when the lecture was or what it was about. I can't. (But I'm sure that it was not about eugenics.). After the talk, I waited until the crowd around him had died down. Then I went up to introduce myself and thank him for creating the business I had already grown so fond of -- the semiconductor business. When you shook his hand, you could tell that he had a huge personality. Clearly physically fit. Strong grip. Piercing, direct eye contact. The typical signs of an alpha dog. I was seven inches taller, 50 pounds heavier, and 35 years younger than Shockley, but he still intimidated me. He was the boss!!

Without question, he was the greatest semiconductor physicist of his time. Unfortunately for him, though, he couldn't stick to physics. He took up eugenics. He preached that the world was suffering from intellectual regression to the mean. By doing so, he became a hated man. His views were, in short form, that too many babies were being produced by parents with inferior intellect and not enough by parents of superior intellect. Luckily (or should I say perversely?) he thought he knew how to solve that problem. He teamed up with a millionaire eugenicist named Robert Graham to open what was referred to as the Genius Sperm Bank. Yes. It really existed. Officially it was "The Repository for Germinal Choice."

There are many different rumors about the sperm bank, but the most prevalent is that you had to be a Nobel Prize winner in order to be allowed to donate sperm to the Repository. In any event, it's certain that Shockley was a donor but uncertain if there were donors other than Shockley. Rumors have it that Shockley was the only donor. It was clear to Shockley that, once discerning ladies learned that they could choose to bear the child of a Nobel Prize winning genius, they would flock to the Repository. Bingo! Smart babies! The regression to the mean problem would be solved! There was just one problem. They didn't flock. Nobody came.

Eugenics? A bad idea!!! The world rapidly turned against him. He became an outcast, even from his fellow professors at Stanford. He died of prostate cancer in 1989. An outcast. Deemed a racist. A pariah.

·········

But the Traitorous Eight lived on.

Episode 3

Part 2

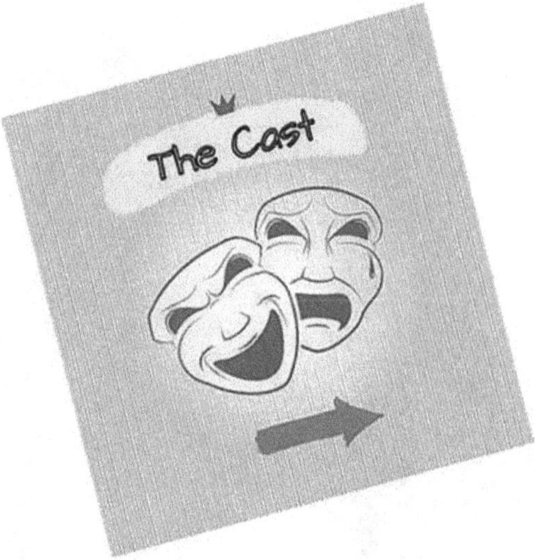
The Cast

The Cast

William Shockley

Image 9

William Shockley was an American physicist and inventor. In 1956 he and two colleagues were awarded the Nobel Prize in Physics for their discovery of the point contact transistor. Shockley formed a start-up company with the goal of manufacturing silicon transistors and silicon four-layer diodes – Shockley Semiconductor. It was the first silicon company in what would eventually be known as Silicon Valley. He hired only the best and the brightest scientists. Then, he proceeded to treat them like little children. That management style doesn't work well with anyone, but it's a particularly bad style to use on an MIT or Cal Tech PhD. Given that prospective employees didn't know what his management style would be, Shockley had no trouble hiring people as he was probably the best-known man in the scientific community in those days. Robert Noyce (more about him is coming soon) said that Shockley somehow got his (Noyce's) phone number and called him one day. Shockley said, "Hello Bob? Shockley here." Noyce said he felt like he was talking to God. He took the job!!

Shockley was quite possibly the smartest man to ever live but, unfortunately, possibly also the most difficult man. Those are both understatements, but I'm not proficient enough with the English language to describe this as it should be described.

John Bardeen, William Shockley and Walter Brattain

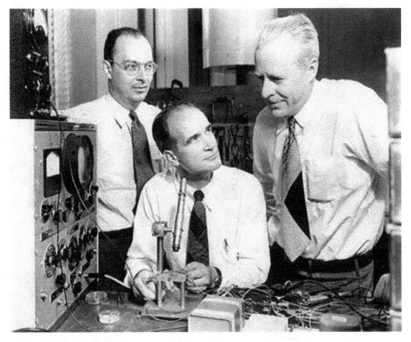

Image 10

The Inventors of the Transistor. From the left: John Bardeen, William Shockley, and Walter Brattain. Shockley was the team leader of the group. All three were PhDs working at Bell Telephone Laboratories in New Jersey when they made the invention. There is no doubt that Shockley was a brilliant man, but it's generally believed that Bardeen and Brattain were the ones who were most responsible for the invention. The three received the Nobel Prize in 1956. Bardeen went on to receive a second Nobel Prize in physics in 1972 for his work on superconductors. He is the only person to ever receive two physics Nobel Prizes.

The Nobel Prize Celebration

Nobel Prize Winner Shockley Hires Young Minds to Perfect His Invention

Image 11

William Shockley was a difficult man to work for. He treated his employees like little kids. Treating your employees like children is frowned on by all the management gurus that I know. But -- at least at first - it was tolerable to the men who worked for him. (You're right. Sometimes people do act like little kids but treating them that way is still discouraged). And then --- the straw that broke the camel's back.

The Nobel Prize (Actually prizes) are the most prestigious awards for intellectual achievement in the world. There are five given every year. One each for physics, chemistry, literature, economics, and peace. A scientist who wins a Nobel prize will typically view that award as the highest point in his or her career. A Nobel Prize is a big, big deal!!

Shockley was technically brilliant, but his manner of dealing with people as well as many of his views were dreadful! In 1956 Shockley (Along with Brattain and Bardeen) won the Nobel Prize in physics. It went to his head. Sadly, instead of making the management style problem better, it made it worse. Shockley began to spend large amounts of time outside of the office giving speeches, being interviewed, and otherwise reaping the ego-rewards of the prize. He became intolerable to many on the days he was in the office. That set the "Traitorous Eight" in motion.

The picture above shows Shockley and some of his employees celebrating the night he was informed that he'd won the Nobel Prize. Shockley is seated at the head of the table. Gordon Moore, Bob Noyce, and Jay Last were there. See if you can spot them.

Little did they know.

Note. In 2020 there were two Nobel Prizes in Chemistry instead of the normal one in chemistry and one in physics. Both chemistry awards were given to women. Not to men. Ladies, you're getting there!!!!

Episode 4

Part 1

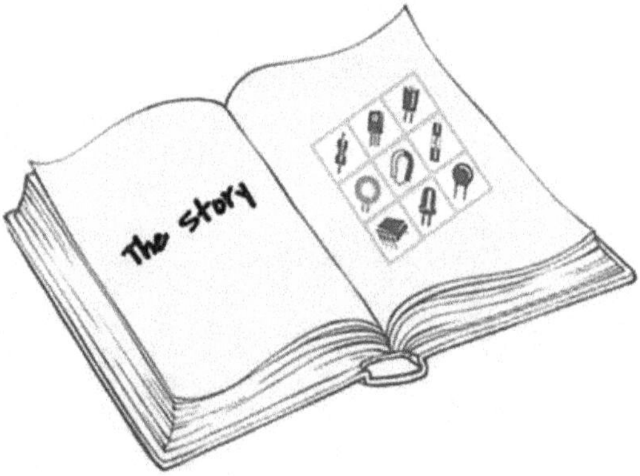
The story

The Integrated Circuit

Noyce and the rest of the Traitorous Eight left Shockley without a clue as to what they would do next. They believed in semiconductors and knew that they were the very best semiconductor guys in the world. Their hope was to find a company who would hire them en masse. After some false starts, Noyce was introduced to Sherman Fairchild. Fairchild was a scientist/engineer who had turned entrepreneur. Among the many companies he had started was Fairchild Camera. The company was very successful: during World War II more than 90% of the military reconnaissance cameras bought by the American forces were made by Fairchild Camera. Sherman Fairchild saw the potential of semiconductors. He and Noyce put together a deal. In 1957 Fairchild opened a new business segment: Fairchild Semiconductor. It was located in Palo Alto, populated initially only by the Traitorous Eight, and Bob Noyce soon became its General Manager. Bob worked for a man named John Carter, Fairchild's CEO, who was based in the corporate headquarters on Long Island, New York. Then--- the event that changed the world. In 1959, Noyce invented the integrated circuit. At just about the same time, Jack Kilby, a Texas

Instruments engineer also invented the integrated circuit. This led to some interesting times.

I knew both Bob and Jack. They were probably the two nicest men I have ever met. Over the years, they were quite complementary about each other. Jack Kilby was as kind and gentle a man as you will ever meet. That was a good thing because he was huge. I'd guess six feet eight or six nine with a big frame. When I shook his hand, I felt as though he could crush mine if he chose to. Luckily, he didn't. Bob Noyce was the greatest "people" guy that you will ever meet. After one meeting with him, you walked away feeling like you were his best friend. He had tons of charisma.

Kilby's patent specified putting all the components on the same piece of germanium (not silicon), but he interconnected them with wire bonding techniques. Obviously, this was totally unpractical with respect to making a real IC. But Bob Noyce got it right. His patent put all the components on the same silicon chip and interconnected them with deposited, patterned, and etched metal --- just the way we still make them today. The Noyce patent is displayed on a plaque in front of the 844 E. Charleston Road building in Palo Alto where he made the invention. Kilby beat Noyce to the punch (and to the patent office) by six months, but you couldn't make an IC without Noyce's interconnect method. After a big legal battle, TI and

Fairchild cross licensed their patents and that was the end of it except for the ongoing arguments over which state should get the credit --- Texas or California. Since I was born in Texas and live in California, I feel confident that, if this thing is ever really settled, I'll come out on the winning side.

While I'm at it, a little bit about Moore's law. Everybody knows about Moore's Law, but here's some perspective from 1965.

When Gordon Moore first articulated his "law," he was predicting that the number of components on a chip would double from a starting point of 32 components in 1964 to a gargantuan 64 components in 1965. The logarithmic graph that he published was so bold as to predict a time when there would be 64 *thousand* components on a chip. To the guys in the fab trying to make these things, that seemed crazy!! I know it seemed crazy to me when I got there in 1968! Gordon firmly believed in the 64K number, but if you had asked him about 64 million, he probably would have thought you were nuts. Today we're doing on the order of 64 billion. Wow. Sounds like more Alice in Wonderland stuff, doesn't it? Fairchild moved to a bigger facility in Mountain View in 1960, but the Charleston Road building is still there. It's a Historical Landmark. The new facility comprised a headquarters building at 313 Fairchild Drive and an adjacent manufacturing building at 545

Whisman Road in Mountain View. Later they added the "Rusty Bucket" (Even though it was a new building, the metal framed structure looked like it was old and rusty) on Ellis Street. When I reported to work on September 9, 1968, I headed to 545 Whisman Road where I spent the next eight years.

Eventually, working for John Carter drove Noyce nuts. Carter didn't see the potential of semiconductors. He didn't want to spend the money to do the job right. He had an east coast management style which was vastly different from the style that had developed in the Bay Area semiconductor business. He didn't believe in stock options for the rank and file. This friction and other issues eventually led to Carter's resignation, but Sherman Fairchild decided not to give the top job to Noyce, who thought he deserved it.

Noyce and Moore decided to leave. Andy Grove, who was running part of Fairchild's Research and Development Labs in Palo Alto, heard about it and asked to join them. (More about Andy later) They left in June 1968. I had accepted my job in May right before they left. Hence, all the craziness in my *"Day One"* and *"Off with their Heads"* chapters. In order to fill Noyce's position, Sherman Fairchild recruited Lester Hogan. Les had been running the Semiconductor Division of Motorola.

In a bit of a coincidence, earlier in his career Hogan had worked under Bill Shockley at Bell Labs. Hogan agreed to join, but only if he could bring eight of his best people along with him. Fairchild agreed. In they came.

And the stuff hit the fan!!

Episode 4

Part 2

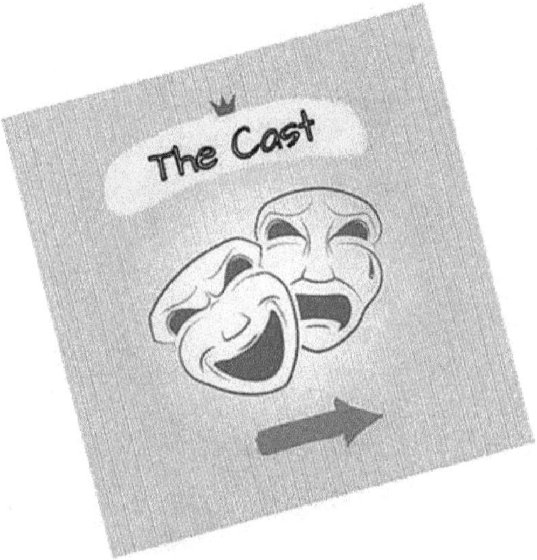
The Cast

The Cast

Robert Noyce

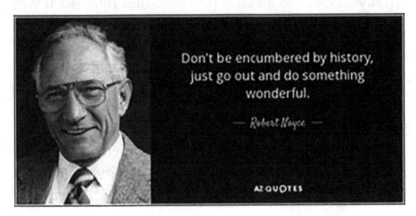

Don't be encumbered by history, just go out and do something wonderful.

— *Robert Noyce* —

AZ QUOTES

Image 12

Robert Noyce received a PhD in physics from MIT. He started his career working at Philco in Philadelphia but was one of the very first hires of William Shockley at Shockley's new company, Shockley Semiconductor. Noyce was brilliant as were all of Shockley's hires. The staff at Shockley Semiconductor comprised what might have been the most brilliant team ever assembled. That team was in complete agreement on two things: **1.** Shockley was even smarter than any of them. **2.** Shockley was the most difficult man to get along with that they had ever met.

Shockley treated his employees like small children. They hated it. Eventually several of them decided that they had had enough -- they were going to quit. They wanted to offer themselves as a team to another semi-conductor company where they would enjoy work more. But – they thought they needed a leader, and they didn't have one. Bob Noyce was a natural leader. That was clear to all who had ever met him. (As it was clear to me when I met him many years later) Noyce hadn't yet decided to leave, but he understood their points of view. Shockley was indeed difficult. They convinced Noyce to join. In 1957 they left en masse - the Traitorous Eight.

In 1959 Noyce invented the first practical integrated circuit. Jack Kilby invented an integrated circuit in 1958, but Kilby's wasn't practical. The fun began!!

Jack Kilby

Image 13

Jack Kilby received a master's degree in Electrical Engineering from the University of Wisconsin. He then began a long career at Texas Instruments. Jack was the first to conceptualize what we now call an integrated circuit, beating Robert Noyce to the patent office by six months. However, Jack's IC was based on an interconnection concept that didn't allow mass manufacture whereas the ICs made today, 60 years later, are still made in the fashion detailed by Noyce in his original patent. In the end, Noyce won the patent battle but Noyce and Kilby agreed that they should share equal credit and both of them did that until their death. In 2000 Kilby received the Nobel Prize in Physics for his invention. Noyce didn't share in the award because the Nobel Prize organization had a firm policy not to award the prize posthumously and Noyce had passed away in 1990.

I first met Jack in Beverly Hills prior to a Wyle board meeting. (Wyle was an electronics distribution company) Jack was on their board of directors and I, as a supplier to Wyle, was making a presentation to their board later that day. We happened to be alone in a conference room. We shook hands and then Jack immediately asked me to tell him about myself -- who was I and what was I doing? I can't remember how I answered, but I remember clearly what I was thinking. "Jack, I'm nobody and I've done nothing. You, along with Noyce, have made the world a better place. Anything I might have accomplished or acquired in life I owe in part to you. How in the world could you want to know about me?" But he did.

That's who he was.

Jack Kilby and Robert Noyce

Image 14

Bob Noyce and Jack Kilby were two of the smartest people that I ever met. They were also two of the nicest. The public debate over who really invented the integrated circuit goes on even today. It will probably never be settled in some people's minds. Certainly, they both deserve some credit, and in fact, both were willing to give plenty of credit to the other guy. Their respective companies, however, were not. Litigation between Fairchild and Texas Instruments for the rights to the integrated circuit patents went on for years.

There were appeals in the patent office as well as in federal courts. It was eventually settled in 1969 with the final nod being given to Noyce. However, by then it didn't matter. Fairchild and TI were both rightfully concerned about what might happen if they were to lose. Consequently, they reached an agreement before the verdict of the final appeal. Regardless of who won, they agreed to cross-license each other and that both companies could assert whatever patent they might end up with against third party infringers – i.e., against the rest of the world. For many years Fairchild struggled to make a profit selling integrated circuits. They lived off the royalties coming from the Noyce patent and other patents they had amassed.

I love this picture. There was no photoshop back then. Jack wasn't standing on a box. Bob wasn't standing in a hole. As you can see, Bob was an average sized man.

Jack was huge!!

Episode 5

Part 1

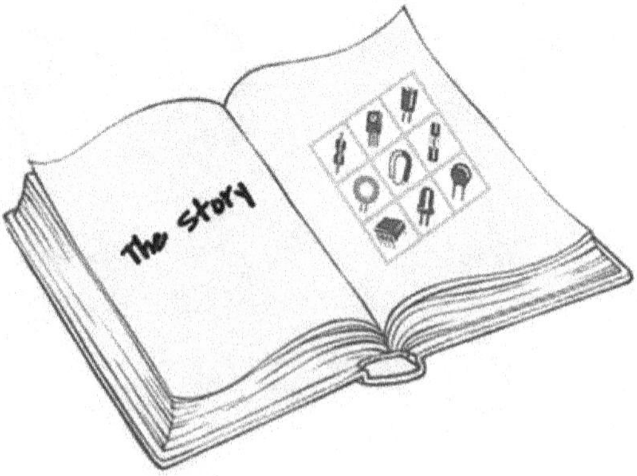
The story

Let There Be RAM

Intel was founded in 1968 by Robert Noyce and Gordon Moore who had left Fairchild earlier that year. They immediately hired Andy Grove. Noyce, Moore, and Grove were a study in contrasts. I had various dealings over the years with Noyce and Grove but have met Moore only twice. They had some things in common but were very different in others. The similarities? Education and IQ. They were all very, very smart and all had PhDs from the very best universities: Noyce from MIT, Moore from Cal Tech, and Grove from Berkeley (His last year there was my first, so we overlapped but I didn't meet him until later in life).

With respect to personalities, there were differences!! Noyce and Moore? – the nicest two guys in the world. Nearly anyone you'd ask would tell you that. In fact, maybe they were too nice? Andy once told me that he thought so. But not many people would say that about Andy. Andy was not "too nice." Most would say he was the toughest, most direct, most in your face guy in the world. Most would say that he had no taste for incompetence and an extremely high standard for what comprised competence. And all of those he found incompetent (Even if only temporarily) paid the price!!! Noyce and Moore founded Intel. The fact that they chose Grove to join them as the third employee might say a lot about their ability to recognize their own strengths and weaknesses. The three rotated through the CEO job. Noyce held the reins from 1968 to 1979 and then passed them to Moore. Moore, in turn, passed them to Grove in 1987. Grove passed them to Craig Barrett in 1998.

Fairchild's management was curious!! The question back at Fairchild was: What was Intel up to? What products were they working on? No one knew. They kept it a close secret. Some rumors had it that they were working on advanced TTL (Transistor – transistor logic) products. Others had them going into the analog space. After all, those two markets comprised the bulk of the IC business in those days. Not so fast! Noyce and Moore were smart guys. They realized that the Achilles heel of computation in those days was memory. The existing memory techniques were dreadful. Almost all memory functions were achieved in those days using core memory. Core memories were made by taking huge arrays of small iron cores -- bits of iron shaped like doughnuts but much, much smaller -- and stringing three wires going in three different directions through each of the cores. What are the most common adjectives used to describe them? Heavy. Bulky. Slow.

It was generally believed that, in order to replace magnetic cores, you would need to be able to sell semiconductor memory for around one cent per bit – about the price of core memory in those days. In truth, though, it was tougher than that. The core memory manufacturers were getting better and better so one tenth of a cent per bit was really a better target. One tenth of a cent per bit? Would that be easy? Or next to impossible? In 1969 I was a product engineer at Fairchild. One of my products was the 9033 – a 16-bit bipolar memory. That's not a typo. Sixteen bits. Two bytes. Except – it wasn't a very useful memory because it didn't have any address decoding. The word lines and bit lines came directly out to the package pins, so the decoding had to be done externally. From what I remember (And admittedly my memory is really, really sketchy), the yield wasn't very good so the die cost might have been around one dollar. Add in the cost of the decoding that should have been on the chip but wasn't and you'd probably be at around a two or three dollar die cost. Adding in the cost of packaging, testing, etc. and then a decent profit margin, my guess is that a hypothetical useful 16-bit bipolar memory from Fairchild would have sold for in the neighborhood of $10. $10 for 16 bits comes to about sixty cents per bit. About 100 times more than the market demanded. So - it seemed hopeless.

Hopeless? That's what Noyce and Moore loved. That's what they were good at. In 1969 they announced their first product. A memory. Not a TTL logic chip. Not an analog chip. A 64-bit memory. A short while later they announced the Intel 1101 --- a 256-bit static random access memory (i.e. an SRAM *pronounced: "ess-ram"*) designed using PMOS silicon gate technology ---a technology that had been a focus at Fairchild when Noyce and Moore left. It was slow – 1 microsecond access time --- and needed some awkward power supplies to make it work -- +5V, -7V, and -10V --- so it wasn't a world beater, but it was a start. Did they make the .1 cents / bit price point? No. Not even close. But they started the ball rolling. Soon nearly every semiconductor company jumped on the RAM bandwagon. Competition and innovation were fierce.

By the time the mid-seventies rolled around, core memory was a thing of the past.

Episode 5

Part 2

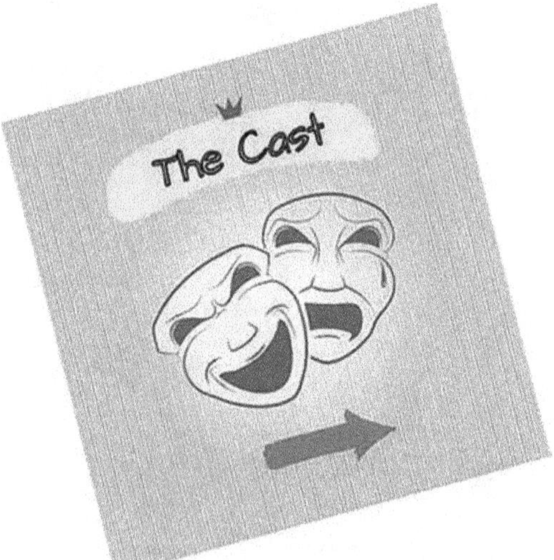
The Cast

The Cast

Gordon Moore

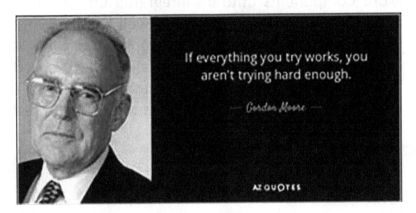

If everything you try works, you aren't trying hard enough.

— Gordon Moore —

AZ QUOTES

Image 15

Gordon Moore was born in San Francisco, California. He attended Sequoia High School in Redwood City. Initially, he went to San Jose State College. Later he transferred to the University of California, Berkeley, where he received a bachelor's degree in chemistry in 1950. Then Moore enrolled at the California Institute of Technology. While at Caltech, Moore minored in physics and received a PhD in chemistry in 1954. Soon after that he accepted a job working for Shockley.

Gordon was one of the "traitorous eight" who left Shockley to form Fairchild Semiconductor. He was Bob Noyce's closest confidant during the Fairchild years. Noyce was the one who came up with the concept of an integrated circuit, but like most of his ideas, he bounced it off of Gordon to see what Gordon thought. During most of his stay at Fairchild, Moore ran Fairchild's R&D facility in Palo Alto. It was while he was running R&D that he published the paper "Cramming More Components onto an Integrated Circuit." That article ended up being incredibly well known. Why? In that article Moore included a graph which today is known as Moore's law. In 1968, he and Bob Noyce left Fairchild to form Intel which later grew into the world's largest semiconductor company. As they say, the rest is history.

Gordon Moore, Bob Noyce and Andy Grove

Image 16

From the left: Gordon Moore, Bob Noyce, and Andy Grove. In early 1968 Noyce was running Fairchild. Moore and Grove were the two most critical people at Fairchild's R&D Facility. They recognized that the Achilles heel of computation in those days was memory. They saw a way to address that problem. I didn't know a lot about semiconductors when I was in college, but I knew that Noyce had invented the integrated circuit and that Grove had written the text on integrated circuits that was being used at Berkeley at the time. When

I accepted the Fairchild job offer in May of 1968, I was elated that I would get to work with such an impressive team. Sadly, by the time I actually started work in September, they had all left and gone to Intel. At Intel they changed the world.

I ended up dealing with all of them later in life.

Episode 6

Part 1

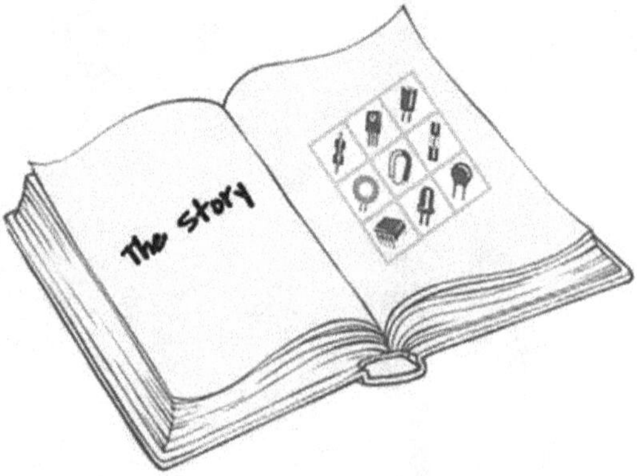
The story

The Story

The Year the World Changed

From time to time I present the History of *Silicon Valley the way I Saw It* to various audiences. I always enjoy doing that. I've learned that the part that audiences like the most is often the Apple/Steve Jobs story. That's not hard to understand. Steve Jobs was truly fascinating! The story that captivates me, though, is this one. I was working at Fairchild in 1971 watching this story unfold. It took me 20 years to understand how important it was. Looking back at it, it still amazes me!!

The years following the Intel 1101 SRAM introduction were dramatic!! In 1970 Intel introduced the 1103 -- the world's first commercially successful DRAM. (Dynamic Random-Access Memory) The yields were very poor. They couldn't ship many. But the concept was born. The stage was set for the future. Then came 1971.

In 1971 the 1103 yields improved, and they were able to ship large quantities. (In fact, by the time 1973 had ended, the 1103 was the biggest selling chip in the world.) Also, in 1971 Intel announced an erasable floating gate memory -- the 1702. The 1702 was a 2K PROM (two

67

thousand bit Programmable Read Only Memory) which stored its contents on the gates of "floating gate" transistors. This by itself wasn't particularly exciting, but there was a twist: By radiating the die with ultraviolet light (Made possible by a see-through lid on the top side of the package), you were able to erase the contents of the memory and then reprogram it. It wasn't a PROM after all. It was the world's first commercially successful EPROM (*Erasable* Programmable Read Only Memory). One of the people on the Intel floating gate memory team, Eli Harari, continued to work for years on the technology helping it get to where it is today – *flash memory*. To oversimplify a bit, today's flash memory is the grandchild of that original EPROM product.

Finally, in 1971, spurred by the creative work of Federico Faggin and Ted Hoff, Intel introduced a microprocessor. Their first microprocessor, the 4004, was only a 4-bit processor but again, it set the stage. And oh - by the way --- they went public in 1971 making many of them multi-millionaires. They couldn't be more deserving!

So, in 1971 Intel commercialized the first DRAM, introduced the first commercially successful floating gate memory, and introduced the first microprocessor. How impressive is that?

Today – 50 years later - those three categories, along with SRAM, still dominate the world's semiconductor market. I would guess that more than 90% of the transistors in one of today's ten billion transistor mega-chips are used in one of those functions. I think the most important innovation in high tech was the transistor. Next – the integrated circuit. Those are more or less no-brainers. My third place? Intel's 1971. That's when the building blocks still being used today were first well defined and commercialized.*

Boy. Did that ever aggravate us at Fairchild. We felt like we were working hard, but they were killing us! How in the world did they do it? How did they invent their way into stardom while at Fairchild we were spinning our wheels? I've met Gordon Moore only twice -- the first time was at a party at Larry Sonsini's home (Wilson-Sonsini is the dominant Silicon Valley law firm). Sometime later I sat next to Gordon at some sort of industry dinner. I can't remember what the dinner was for or why I sat next to him, but I do remember the conversation. I asked him that question: "How did you do it? How did you build and incentivize a team to foster such innovation?"

His answer was, "First - you hire really smart people. Second - you make it clear that they will get the credit for their work. And third – you let them know what you wish they would invent.

If you don't do that, lots of crazy things are going to get invented that you will have no use for." I don't think that Gordon could pick me out of a police lineup if he saw me today. (I hope he never has the occasion to do that!!) We met only those two times. But I've heard so much about him from some of my friends who worked for him I feel like I know him. Gordon was always known for being modest and unassuming. He famously drove an old car to work every day so people wouldn't think he was putting on airs. He wanted to be seen as a "regular guy." Regular guy? For a guy with an IQ of 200 and several billion dollars in the bank??? Tough to pull off, I suspect. (I wish I had first-hand knowledge.) But -- if you sit next to him at dinner, that's the feeling you get. By the way, Gordon is at least five billion dollars lighter in the wallet than he used to be. He once donated 175 million shares of Intel to charity. And he's not done giving. The Gordon and Betty Moore Foundation donates billions to charitable causes focusing on ecology and scientific innovation.

The Intel story isn't all glory. Real life stories never are. Their worst stretch came in the early to mid-eighties. Japan Inc (The four biggest IC manufacturers in Japan were often referred to as Japan Inc: Fujitsu, Toshiba, NEC, and Hitachi) had embarked on a plan to conquer the world's integrated circuit business by taking control of the three M's: Memories, Microprocessors, and Master slices (gate arrays). They (the Japanese) were already good at manufacturing. They already had low costs. Their strategy was to bomb prices until the American companies, (who were less well capitalized and had huge pressure from shareholders to keep earnings high), gave up and went out of the business. The easiest and most obvious target was the DRAM market. DRAM prices were dropping like a rock. Intel began losing money for the first time since their IPO in 1971 (Initial Public Offering is the technical term for "going public"). DRAMs, which once represented 90% + of Intel's revenues, were down to a few percent by 1984. Even though Intel had invented the DRAM, by 1984 they had no particular strategic or technical advantage. They were good at DRAMs. So was everybody else.

On the other hand, because of their design win in the IBM PC, they had attained a near dominant position in the microprocessor market. That market had barriers to entry much higher than DRAMs ever had or could have. Should they exit the DRAM business? There were great debates. A quote from someone inside the company went, "Intel leaving the DRAM business would be like Ford leaving the auto business." In the end they opted to exit. The P&L was messy for a couple of years.

A quote from their 1986 Annual Report went, "We're pleased to report that 1986 is over." Then, their shares grew steadily until their Market Cap hit nearly $300 billion. ("Market Cap" is the value of a company as set by the stock market. The share price multiplied by the number of shares outstanding equals the Market Cap). I guess it was the right decision. When I was young, America was better at everything. Cars, steel, shoes, clothing, engineering. Everything!! "Made in Japan" was a derogatory term. China didn't matter at all in the world economy. Taiwan and Korea were trying to matter, but they didn't. Today, as a nation, the USA has lost a lot of that.

Industries that used to pay for our way of life are now struggling in America. But -- we're still really strong in high tech. Sure, we have competition, but we're a force to be reckoned with. In fact, *the* force to be reckoned with. If there were a king of high tech, it would be the USA. Without the contributions of Intel (And Microsoft) that wouldn't be true.

Thanks Gordon. Thanks Bill Gates.

Disclaimer: Inventions are rarely made out of the blue by one person or one company. Yet it's usually an individual person or company that ends up getting credit in the history books. This episode is a perfect example. The concepts of SRAMS, DRAMs, EPROMs, and microprocessors didn't necessarily originate at Intel. In fact, there were probably dozens of companies and hundreds of people that in one way or the other theorized about or worked on SRAMs, DRAMs, EPROMS, and microprocessors. Still, the race goes to the swiftest. The people at Intel were the ones who put it all together, introduced real products, marketed those products, and got real, significant sales from them. As a long time AMDer, it may pain me a little to admit it, but Intel deserves a huge amount of credit.

Episode 6

Part 2

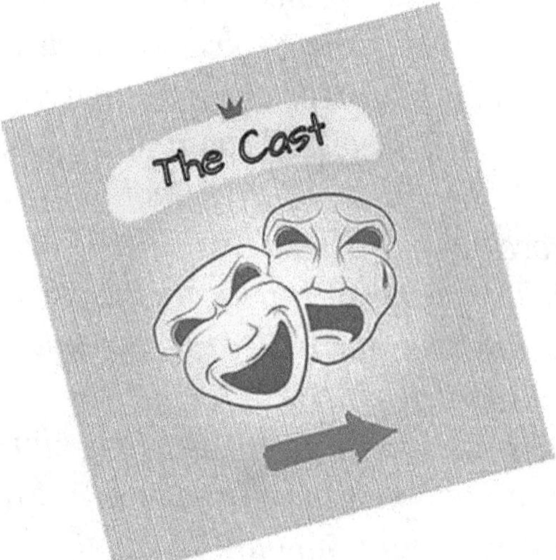
The Cast

The Cast

The Microprocessor

Image 17

By the way, what the heck is a microprocessor anyway? What does it take for a human being to "think" about something. **First**, the ability to reason: to take inputs and draw conclusions from them. A simple example? Touching that hot stove hurts my hand. **Second**, the ability to remember things. An example? What's your phone number? Computers work the same way. The "reasoning" part of a computer resides in something called an ALU (Arithmetic Logic Unit). And the part of the computer that remembers things? Duh. .... That's called memory. There are three types of circuits that combine to handle the "memory" aspects of computers: SRAM, DRAM, and Floating Gate (In today's embodiment floating gate is called "Flash').

The Intel business plan and early Intel products addressed these memory functions. Their first big winner was an SRAM. Then came a DRAM. Finally, in 1971 came what they referred to as an EPROM. An Electrically Programmable Read Only Memory. That's a mouthful. What does it mean? What's it good for? SRAMs and DRAMs have a drawback --- They're paid to remember things and they do a good job of it, but when the power is turned off, they forget everything! They are both "volatile" types of memory. Sometimes you don't want that. Who wants a music player that forgets the songs you've downloaded every time you turn it off? Who wants a phone that forgets your contact information every time you turn it off? The Intel EPROM remembered when the power was turned off!! The EPROM was "non-volatile." The technology has advanced a lot since 1971. It's no longer called EPROM technology. Now it's called FLASH. Today's devices are approaching a billion times the density of Intel's first 1702 product which held 256 bytes (2048 bits). By today's standards that product would be nearly useless -- one song on your iPhone uses around 4 million bytes.

A very basic microprocessor puts an ALU, some memory, and some complex circuitry that tells the ALU what to do and when to do it onto the same chip along with some software. That software is called the "micro-code." It's stored in the non-volatile memory.

Because it's stored in non-volatile memory, it's not called software. Instead, it's called "firmware." The first commercially successful microprocessor was the Intel 4004, which was announced in 1971. The 4004 was developed by an engineering team staffed by all-star engineer/scientists Federico Faggin, Stanley Mazor and Ted Hoff. By today's standards the 4004 would be regarded as a toy. It was millions of times less powerful than today's processors. Still, the longest journey starts with a single step.

(Warning: the above "technical" descriptions are grossly oversimplified! The engineers who read this will probably file a class action lawsuit against me, but a real explanation would go far beyond the scope of this book.)

Federico Faggin

Image 18

Federico Faggin received a PhD from the University of Padua in Italy. In 1968 he moved to California and took a job at Fairchild's R&D facility in Palo Alto. I met him shortly after that when I needed a good way to etch silicon nitride. (In those days nitride was notoriously hard to etch). Federico had figured out the best way to do that. The people at R&D affectionately referred to his answer to the problem as "Freddy's etch." Freddy's etch had seemingly every kind of acid and otherwise evil stuff in it. It was scary just to look at it.

At Fairchild he worked mostly on processing --- the steps taken in the wafer fab required to make a wafer. Federico moved on to Intel where he was more of a circuit design guy - what's the circuit schematic required to implement a certain function and what does that look like when you translate it into a mask? (Masks are the mid-point between the description of what a product must do and the complex chemicals and equipment used in the fab to actually make a wafer.) Federico was the key figure in the development of the first commercially successful microprocessor - the 4004. After Intel, Federico became a serial entrepreneur. He was the co-founder of Zilog and then the founder/CEO of both Cygnet and Synaptics.

Ted Hoff and Stanley Mazor also deserve credit for the 4004, but I've never met Ted or Stanley. Since I'm limiting this book to people with whom I've met, I haven't included Ted or Stanley in the "cast."

Eli Harari

Image 19

Eli Harari received a PhD from Princeton in 1973. He then worked at Honeywell, Hughes, and Intel focusing on improving Floating Gate technology (The Intel 1702 EPROM was the first commercially successful example of Floating Gate technology.) The technical gap between Intel's 1702 and the massive flash chips of today is immense. Eli was probably the biggest contributor to filling in that gap. He is generally credited with being the inventor of EEPROM - one of the most significant steps between the 1702 and the huge flash chips of today. In 1988 he founded SanDisk (Originally called SunDisk) and was the SanDisk CEO until he retired in 2010.

In 1991 SanDisk developed and announced the first commercially successful SSD (solid-state drive). Prior to then, hard disk drives were required if you wanted to store large amounts of data – something you would need to do if you wanted to store music or photographs. Hard disk drives were relatively bulky and expensive. The idea of an SSD was to use flash memory instead of a disk drive but to do it with the same "form factor". That is, to make it in the same size and shape as the disk drive that it would replace. SSDs have revolutionized the world. Essentially all cellphones and digital cameras today as well as most computers are based on SSD technology.

Episode 7

Part 1

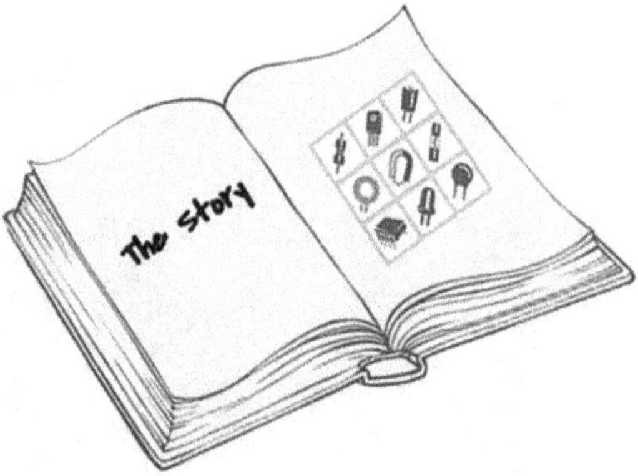
The story

Layoffs a la Fairchild

You wouldn't think that layoffs would be a subject that I'd want to talk about in my stories of "*Silicon Valley the Way I Saw It.*" The subject is just plain distasteful!! Still -- layoffs were a major part of the Valley's culture back in the day. To give a true picture of how it felt to work at Fairchild in the early 70s, this story must be told.

When Les Hogan and his heroes joined Fairchild, the company had been losing money. As I look back on it, I imagine they might have been in the middle of a cash crisis. Also, in looking back it seems that there would not have been a quick fix. TI clearly had a superior cost structure to ours. Our wafer sort yields weren't good at all and we didn't have an inexpensive plastic package to match the one that TI had developed. The picture from the top couldn't have been rosy. One day Les Hogan was interviewed by someone in the financial press. He was asked what he intended to do to stem the losses. His answer, "That's not a problem. We're going to reduce the headcount by a third. That will get our spending back in line."

That quote made the front page of the business section of the local newspaper. It probably felt good to the investors when they read it, but it felt really, really bad to the people who worked there. That was not what we wanted to hear! And that was the beginning of some serious layoffs.

Once they started, it seemed like every Friday somewhere in Fairchild there were layoffs. TGIF didn't apply at Fairchild!! Everyone went straight to the cafeteria Friday mornings. No one bothered going to their desk. Everybody was scared to death. Everybody needed their job and knew that there was a very good chance that they would lose it in the next few minutes. Lots of gallows humor. Lots of camaraderie. Everybody loved everybody else. There are no atheists in foxholes. And then, at 9:00AM, they started.

There was a very well known "journalist" who covered the semiconductor industry in those days. His name was Don Hoefler. I never met Hoefler. I think he had worked at Fairchild at one time and that there was bad blood when he left. Maybe he had been fired? Maybe an "*Off with their heads*" casualty? I don't know. He started writing a weekly industry newsletter. It was always very negative towards Fairchild. Needless to say, the upper echelon at Fairchild were not enamored of Hoefler. I heard several times that anyone caught in possession of one of Hoefler's newsletters was subject to being fired - but that might have been just a rumor. After all, we all read it (being careful not to get caught) but so far as I know, no one was ever fired for that offense.

One of Hoefler's newsletters dealt with *"Layoffs Fairchild Style"* describing how Fairchild employed three unique tactics for implementing layoffs. The paging system layoff, the locked door layoff, and the retroactive layoff. How did those work?

1. The Paging System Lay-off.

This was the standard. I witnessed this one many, many times. ------ It's Friday morning. We're all huddled together in the cafeteria. Around 9AM the paging system cranks up. "Bob Martin 2867." ----- Everybody knew what that meant. Bob Martin (who was a real person and a really delightful guy) knew what it meant too. Bob or some similar victim would stand up and start shaking hands. After he said goodbye to everyone, he'd walk over to the phone and call 2867. 2867 was, of course, the HR department (HR was called "personnel" in those days). "Bob, this is Bill, can you drop by to see me?" That would be the last that anyone would ever see or hear of Bob Martin.

2. The Locked Door Layoff

I never saw this one, but Hoefler swore it happened. I think it may have been used up at the

R&D facility -- that's where the best kept technical secrets resided. There was great fear in those days that company secrets would be stolen by people leaving the company. The Basic Data Handbook was the result of a lot of work that Fairchild rightly didn't want to fall into the hands of the start-ups who were trying to eat Fairchild's lunch. On the other hand, anyone who was leaving for any reason would be tempted to take a copy of it on his way out. How could you keep that from happening?

According to Hoefler, if you were going to lay off someone in possession of a lot of key knowledge, then the way to do it was to have the facilities department change the lock on the victim's door the night before. Then, when the victim arrived in the morning and found his key wouldn't open the door, he'd go to ask his boss what was going on. Big mistake! His boss, of course, would give him the axe. That way he had no pre-warning and couldn't sneak the key information out before the axe fell. Whether or not this really happened, it was commonplace on Thursday afternoons to see people preparing for what might turn out to be a bad Friday morning.

3. The Retroactive Layoff

This happened if you were unlucky enough to be selected for downsizing when you were out on vacation. There were no cellphones then.

It could be tough to contact someone away on vacation. Not a problem! When the victim returned, he was informed that he had been laid off and that there was good news and bad news:

"The good news is we gave you two weeks of severance pay."

"The bad news is you were laid off three weeks ago."

Perversely enough, I don't think it was as bad for the victims as it might seem at first blush. The valley was rife with start-ups and many of them were hiring. I'd venture to guess that most of the victims got jobs that were as good or better in short order and that their careers played out better than they might have if they'd stayed at Fairchild.

This story may seem a bit frivolous -- not to any particular point. But there is a point. It's a snap-shot of how Fairchild was back in the day, and to a lesser extent how the entire semiconductor community was. It was seen by all and consequently shaped the thinking of future generations. Jerry Sanders, for example, watched this unfold and used it to mold some of the management theories that he would later employ. (See episode #11 in this book). Jerry, by the way, once said:

"Being fired by Fairchild was the best thing that ever happened to me."

Episode 7

Part 2

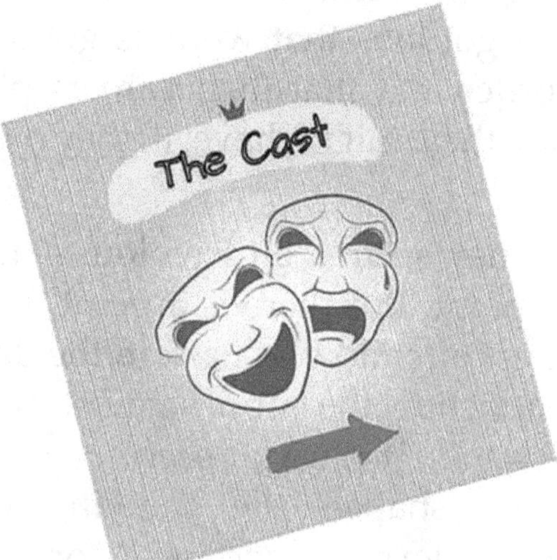
The Cast

The Cast

Fairchild Assembly

Image 20

Women are the mainstay of Fairchild Semiconductor's busy production line which turns out transistors and other electronic gadgets. Because of high labor costs, however, the firm is relocating most routine production operations to other areas of the nation.

This is a shot of the assembly area at 545 Whisman Road in Mountain View. Sexism was alive and well. In those days all of the line operators were women. All of the engineers and managers were men. But -- sexism didn't apply when it came time for layoffs. When times got tough, Hogan declared that he was going to cut the headcount by 1/3. He lived up to his word. Every Friday there was another layoff until the quota had finally been met. Sex? Race? Religion? Didn't matter. Chances were, "You're outta here!" It was no fun!!

Other photos of the day (circa 1965 or so) all look more or less the same. Want proof? Google pictures of the "traitorous eight," or the inventors of the transistor, or the founders of Intel, etc. Not only were nearly all the engineers and inventors Caucasian males, but they were also usually wearing white, long sleeved shirts and dark ties. And the operators? Always women. If you took photos of a similar area today and compared them to shots from 1965, you'd think you were on a different planet. Maybe even a different universe. People who say that things haven't changed are just plain wrong.

Walker's Wagon Wheel

Image 21

I can't remember the last time I had a glass of wine at a business lunch. It was at least 25 years ago. That's a big change from the early days of Fairchild. People drank heavily then, and usually did it at Walker's Wagon Wheel -- a bar/restaurant just down the street from the 313 Fairchild Drive/545 Whisman Road headquarters complex. Friday nights were always busy there, but layoff Friday nights were incredible. All the people who had been let go earlier in the day would head to the Wagon Wheel to say their goodbyes and to argue that it should have been someone else - not them - who got the axe. The place did so much business on those days that if you didn't get there early, you couldn't get in. That wasn't a huge problem for the unfortunate guy who had just lost his job. Layoffs were always done in the morning.

Episode 8

Part 1

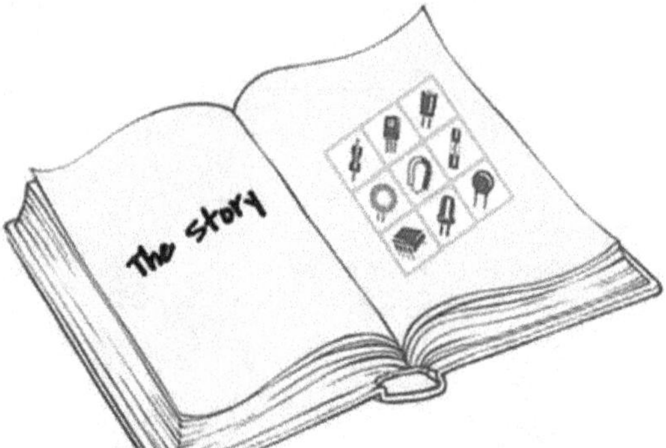
The story

The Story

Texas Instruments & the TTL Wars

Most people in the IC business understand very well that Transistor-Transistor Logic (TTL) products dominated our industry for 30 years or so. They'll also probably know that Texas Instruments (TI) was the king of TTL. But, if you ask those people what TTL is, most won't have any idea. If you're one of those people, rest easy. You're about to find out.

There were three IC companies that really mattered in the early days -- Fairchild, Motorola, and Texas Instruments (TI). There were a handful of wannabe's as well -- National, Signetics, Amelco, Siliconix to name a few -- but Fairch, Mot, and TI were the big guys. Intel had not yet arrived on the scene. The first standard logic family was Fairchild's **RTL**. --- **R**esistor -**T**ransistor **L**ogic. (Inputs came in through **R**esistors. Outputs went out through **T**ransistors. They were **L**ogic devices). It was built using bipolar transistors as was almost everything in those days. Micrologic was the name they gave the family. Sadly, **RTL** was brain dead. Fairchild had a great head start. The Noyce patent really worked. The Kilby patent didn't. But – the RTL circuit that Fairchild chose had big problems -- namely fan in, fan out, noise margin, and speed. Sadly, those pretty much captured everything that mattered in those days. So - Fairchild switched to **DTL**. --- Diode-Transistor Logic. (The inputs came in though **Diodes**. The outputs went out through Transistors. It was Logic). **DTL** had been invented by others earlier. IBM had used discrete versions of it in their 360 series of main-frame computers. (A discrete version is one that uses individual transistors and diodes – it's not an integrated circuit). It was a much better design than **RTL**. It solved all of the **RTL** problems except speed. Fairchild introduced their **DTL** family in 1964.

TECHNO-BABBLE ALERT. If you're not a circuit geek, skip the next paragraph!!

DTL gates included back-to-back diodes. The cathode of one diode (called the input diode) went to the input pad and the cathode of the other diode (called the level setting diode) went to the base of a bipolar transistor that we called the phase splitter. The diode anodes were common. Why not put those two diodes in the same isolation region? Instead of two diodes, use a single bipolar NPN transistor! The input diode would be the emitter-base junction of the transistor. The level setting diode would be the collector – base junction of the transistor. (For a two-input gate, the transistor would have two emitters. For a four-input gate, it would have 4 emitters. That schematic diagram looks a little weird, but it's no problem to make) Why was that better? When the input pad went low, the transistor turned on and yanked the charge off of the base of the phase splitter. That shortened the time required to turn off the phase splitter which was the number one problem facing the engineers who were trying to speed up the circuits. As with most inventions, it isn't always completely clear who deserves the credit. A real, working IC version of a TTL gate had a lot more to it than my simplified description above. Tom Longo, working at Sylvania, was the first to put it all together into a commercially successful IC.

It was a big breakthrough!! After working through a few "pet" names (Longo called it SUHL and James Buie of TRW called his version TCTL), it ended up being called **TTL** (Transistor – Transistor Logic. Whatever you called it, TTL was better! It was faster than DTL with similar costs. Longo and a few others may have invented TTL, but it was Texas Instruments who made hay with it. TI introduced their 5400 TTL family in 1964 --- the same year Fairchild announced their DTL family. But TTL was better! TI soon put their TTL products in an inexpensive plastic package which they had developed. The plastic package gave them lower costs than anyone else. Early on TI sold TTL gates for $1.00 (actually twenty-five cents per gate for the quad, two input NAND gate. There were four gates in each package. Four gates in a package sold for one dollar.) That was an outrageously low price for the times. Bottom-line? TI stole the march. TTL took over the world and TI became the king of that world.

Why didn't Longo et al get the credit they deserved? TI took such a huge, early lead in TTL that everybody thought it was a TI invention. It wasn't. TI recognized a good thing when they saw it and they jumped on it. They pounded the market with it. There's a Harvard case study wrapped up in this somewhere.

And Tom Longo? He later moved to Transitron and then to Fairchild. He was a very smart guy and a butt kicker extraordinaire! I couldn't decide if that was a good thing or a bad thing. I worked several levels below him at Fairchild, so I didn't interface with him on a daily basis. The butt-kicking was sort of fun to watch --- unless it was your butt being kicked. (Of course, now and then I took my turn.) Speaking of butt-kicking, though, Tom never liked the fact that TI was kicking our butts with the product that he had invented.

Fairchild, seeing the success that TTL was experiencing, later tried to get into the TTL market --- at first by introducing a proprietary family of TTL products (the 9000 family) instead of by second sourcing the 5400 family. But there was a problem: we didn't really understand collector – emitter leakage. (Iceo). Two things we knew about Iceo, though, were that it was a bad thing to have and we had plenty of it! Iceo was our dominant yield problem. This was exacerbated by gold doping. The big problem with respect to speed those days was turning off bipolar transistors. Turning them on was easy, but to turn them off you had to wait for all the minority carriers to exit the base region. That took forever. Somebody had figured out earlier that a few gold atoms in the base region would help the minority carrier lifetime problem. We called that gold doping. You had to gold dope in order to make the speed requirements. The problem? Gold made the Iceo problem worse. When we gold doped, our leakages went up, our yields down, and our costs soared.

So --- circa 1969 -- our yields were bad and costs were high. The solution? Some of the Motorola guys brought in by Hogan in turn imported what they thought was the Motorola process (Sadly -- it wasn't quite) and installed it in one of the Mountain View fabs.

The yield was pretty good right out of the chute, so we went into production. We didn't do HTRB (High Temperature Reverse Bias is a simplified form of life testing). There were no formal qualification requirements at Fairchild in those days.

We just went straight into production. Big mistake!!!!! We had screwed up!!! There was an Op Life (reliability) problem* that we would have caught if only we'd done HTRB. Several weeks' worth of circuits from our early production wafers were unreliable! They went bad after being in use for a short period of time. Our customers were outraged. Heads rolled. The red queen was still lurking.

Off with their heads was still operative.

*If you're a grizzled IC engineer, you easily understand all of the above. If you're not – it could look like Greek to you. Here's a translation. Obviously if you sell a good circuit to a customer, he'd like it to stay good -- not to go bad just after his product gets out to the end customer. In IC lingo, a product that stays good is "reliable." One that goes bad is "Not reliable." These days ICs are fantastically reliable. In the old, vacuum tube days, tubes had what is called a wear-out mechanism. You knew that they would go bad after a certain amount of time. A vacuum tube is just a fancy electric light bulb. All incandescent light bulbs eventually go bad – so do all vacuum tubes. There's no wear-out mechanism for a properly made integrated circuit. Unless there's some sort of defect in them, ICs will stay good for a long, long time. That wasn't necessarily the case in 1969.

In order to assure the reliability of our new process we should have done HTRB (High Temperature Reverse Bias) also known as Op Life (Operating Life) testing on a large sample of the products before actually shipping them. That would have meant:

- Heating them up to a temperature far above boiling temperature. (Heat makes a circuit that is going to fail do so much faster.)

- *Powering them up.*
- *Running them for a period of time -- sometimes a week -- Sometimes six weeks.*
- *Testing them. Are they still good?*

We didn't do it. Who screwed up? Well – for one –me. I was the product engineer. But I dodged the bullet. Barely a month earlier I had been a production supervisor – not an engineer. I had just moved into engineering. I had no idea what I was doing yet. I had never heard of HTRB or Op Life. It was a good enough excuse. I lived. My big boss who brought the process in from Motorola didn't.

Off with their heads.

Episode 8

Part 2

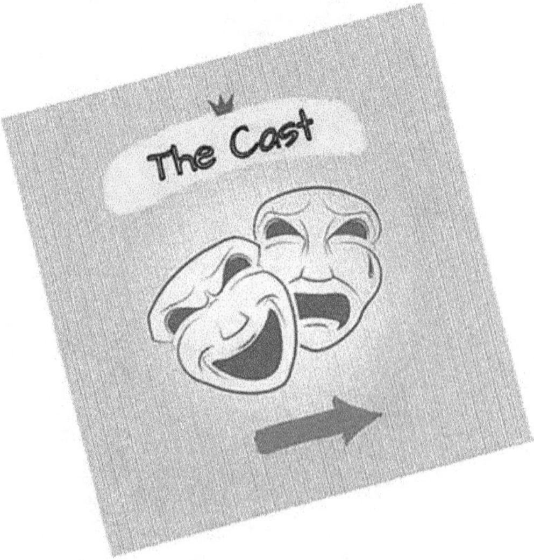
The Cast

The Cast

TTL Gate

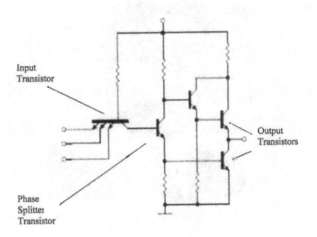

Input
Transistor

Output
Transistors

Phase
Splitter
Transistor

Three Input TTL NAND Gate

Image 22

This is a TTL gate. it seemed complicated at the time, but it's not an issue these days to put a few billion of them (Actually a few billion of their CMOS equivalents) into an integrated circuit. Texas Instruments dominated the IC world for decades selling TTL gates. TI had done a great job of reducing their costs of making TTL gates. During a high-tech recession in 1974/1975, TI sold gates for as low as eight cents per package (That meant two cents per actual gate – there were four gates in each package). At Fairchild, our costs were somewhere around twelve cents per package. Match TI and sell them for eight cents? Or go out of business? We were in trouble.

Tom Longo

Image 23

Tom Longo received his PhD in physics from Purdue University in 1957. He took a job at Sylvania where he developed the first commercially successful TTL circuit. In the very early days of ICs, Fairchild's work was overwhelmingly focused on how to make them. Little thought was put into what to make — specifically what kind of circuit design technique should be employed. Fairchild's first integrated circuit was what was called an RTL design. RTL had numerous problems. Tom's new TTL techniques solved those problems well. He later moved to Transitron and then to Fairchild as the General Manager of Digital Integrated Circuits. He was my "big boss" for about five years. Even though Tom had done the breakthrough engineering work on

TTL, he spent his life chasing Texas Instruments who early on saw the benefits of TTL, acted quickly, and took control of the TTL market thereby managing to dominate it for two or three decades. In 1985 Tom founded Performance Semiconductor where he was the CEO. Tom was a really smart guy but a very demanding boss!

Episode 9

Part 1

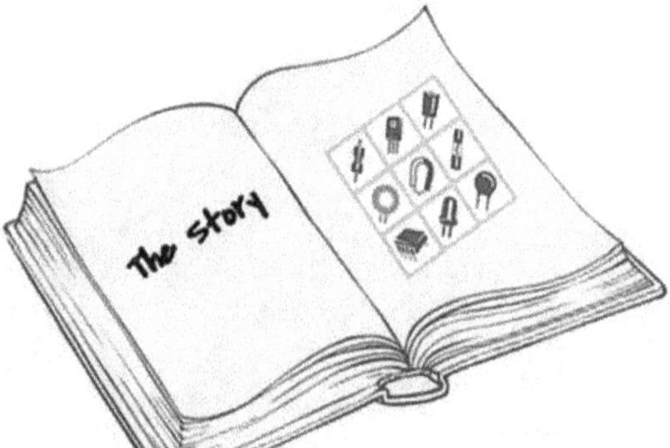
The story

The Story

The P&L Review: Fairchild Style

In 1972 plus or minus a year or two I was working as a supervising engineer in one of the bipolar digital product groups. My boss was a man named Jerry Secrest. He was a great boss – he taught me most of what I knew about ICs in my youth. Jerry had responsibility for a product line. That meant that the Fab, the product and process engineers, and the test area all were under him. It also meant that he had P&L (Profit and Loss) responsibility. That sounds like a good thing, doesn't it? It wasn't!!!

P&L responsibility forces you to open your eyes to how life really is. Sometimes reality isn't fun. It's not hard to find someone today writing that profits are evil and the people who try to make them are even more evil. But reality tells a different story. If you're a private company, to get the company going and keep it going you have to raise money. To do that, you have to tell the potential investors that once you're up and running, you're going to be making nice profits and that the profits will grow over time. If you don't, they won't invest. Going back later and telling them, "I was only kidding" is ill-advised!!

If you're a public company, the same argument holds true, but on steroids! People buy your stock because they think it's going to go up. Why do they think that? Because you led them to believe it. In the long run, stocks go up when earnings go up, so it's your job to make earnings go up. There's no way you could go back to the share-holders and say, "Gee. I changed my mind. Prof-its are evil. I'm not going to try to make any. My plan is to make the stock go down." What, if in-stead of being the CEO, you owned some of that stock? You'd kick the CEO out in a nanosecond. Then you'd sue him --- and you'd win the suit!

Oh. One other thing. In both cases, if you lose money long enough, you'll eventually run out of it. You go broke. So --- it's important to be profit-able!!! Yes. You also have a responsibility to the people who work there and another to the extend-ed community. Trying to take care of all three si-multaneously was the hardest thing I ever did!!!! But --- when the dust settles --- --- it's important to be profitable!!!

Sometimes capitalism sucks -- but I've been in most of the formerly communist countries back when they were just coming out of communism. They were absolute economic disasters. So, cap-italism may well suck, but all the other systems that I'm aware of suck more. Far more!! By the way. I am not affiliated with either political party. They both aggravate me!

Back to the story: Each operation at Fairchild had a regular P&L review with Wilf Corrigan, then a Vice President and later to be Fairchild's CEO. Besides Wilf, his top financial guy and several people from the operation being reviewed attended. I worked in the Digital Integrated Circuits group (DIC) which was run by a guy named John Sussenburger (We called him Suss). Jerry worked under Suss. Suss worked under Wilf. The financial reviews looked at Sussenberger's P&L and its component P&Ls. (Tom Longo, Paul Reagan, and Dave Deardorf were also involved in the organization as high-level managers at various times in those days, but I don't recall any of them being in the meeting)

Fairchild had a thing they called their HiPot List. (High Potential List) You were put on the HiPot list if you were a high potential employee who seemed to have the ability to work your way into upper management. To my great delight, in 1971 or 1972 they added me to that list. Wilf had put in a rule that at each review a different person from the HiPot list should come to get some seasoning. One day Jerry told me that my turn in the barrel had come. It was my chance to go to the review.

I went to the conference room at least 20 minutes before the start time. There was a large potted plant at the back of the conference room. I figured out that one of the chairs in the back was partially hidden from view by that potted plant. Naturally

I took that chair. (I was scared to death of all the Fairchild high-level managers) Then I waited. After a while the real attendees came filing in and the meeting began. Wilf Corrigan is astute! There is no tricking Wilf with the numbers. They would put up very complex foils absolutely full of numbers and Wilf would immediately zero in on the number that made a difference.

He was very direct, but very polite. No screaming, shouting or table pounding even though DIC was losing money. We all understood that Wilf Corrigan didn't like losing money, so the review didn't have a good feel to it - all the attendees were on pins and needles. But – it didn't seem to be a problem. Wilf was calm and cordial. He very calmly went about getting an understanding of what was going on. When all the data had been presented, Sussenberger asked Wilf if he had any questions. I thought, "Wow. This isn't so bad. We're losing money, but Wilf understands. Nobody got beaten up or fired. What was I worried about?"

Wilf said, "Yes. A couple of questions: How much money did you say you were losing?"

(I don't remember the actual numbers, so I'll make some up.)

John: "Oh. We're losing about a million dollars, Wilf."

110

Wilf: "What does your average employee make, John?"

John: "Gee I don't know exactly, but I'll guess about $20,000."

Wilf: --"Hmmm. That comes to 50 people, doesn't it?"

John: "Well, you could look at it that way, Wilf."

Wilf: "I do look at it that way, John."

Wilf: "OK. I don't have any more questions, but John, I'm planning to do you a favor."

John: "What's that, Wilf?"

Wilf: "Tonight I'm going by the hardware store on my way home. I'm going to buy one of those clicker/counters.... you know - the little mechanical things with a button. Each time you hit the button with your thumb it ups the count by one."

John: "Great Wilf! That's great!!! Sounds really good!!! ....I like that!!!..
....
....

… but … what are you going to do with that?"

Wilf: "I'm going to go into your building Monday morning. I'm going to stand in the main hall. Each time someone walks by me I'm going to ask him if he works for John Sussenburger. If he says 'Yes', I'm going to say, 'You're fired' and click the button. When the count gets to 50, you'll be profitable! ----- John, you're really going to enjoy running a profitable business!"

I stayed out of the hall that Monday morning!

Wilf Corrigan went on to become the founder / CEO of LSI Logic

Episode 9

Part 2

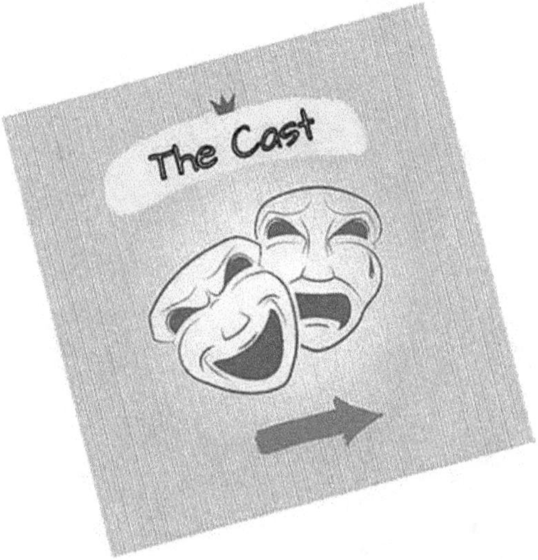
The Cast

The Cast

Wilf Corrigan

Image 24

Wilf Corrigan was born in Liverpool, England. He received a degree in Chemical Engineering from Imperial College in England. He then moved to the United States and took a job at Motorola where, in his early days, he worked as an epitaxial growth engineer. He moved on to Fairchild in 1968. He was one of the "Hogan's Heroes" group of executives that Les Hogan brought with him when he took the reins of CEO of Fairchild. Originally Corrigan was the vice president of the discrete component side of Fairchild. (Fairchild's semiconductor operation was divided into two groups: diodes and transistors were in the discrete group and integrated circuits in the IC group.) Soon he was promoted to be the vice president of the entire semiconductor division thereby taking on the responsibilities of Gene Blanchette, another Hogan's Hero, who had just been let go. *Off with their heads!* In 1974 when Hogan himself was fired; Corrigan took over as the new CEO. In 1979 Fairchild was sold to Schlumberger -- a move that eventually resulted in the demise of Fairchild as we knew it. Corrigan then founded LSI Logic and was the CEO of LSI for 25 years before retiring in 2005.

Episode 10

Part 1

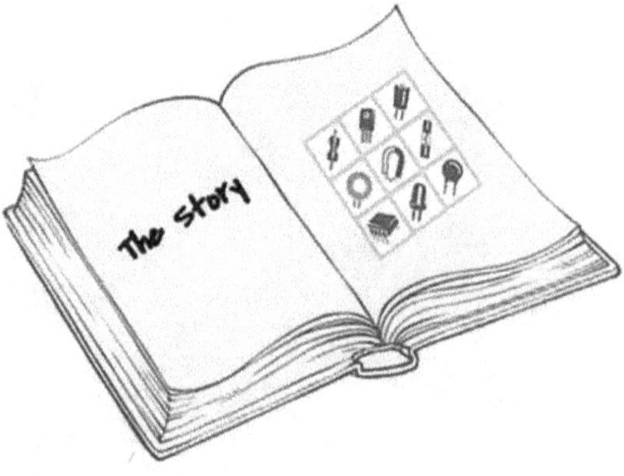
The story

Fairchild's Death March

How did it end for Fairchild? Badly!!!

In 1966 Fairchild was the number one supplier of integrated circuits. That was as it should have been. After all, Fairchild had invented the IC. But in 1967 TI passed them. Still, Fairchild remained a strong #2. By the time that the mid-seventies arrived, though, they were fading. Motorola and some others had passed them in sales by then. Fairchild was clearly struggling and beginning to look like an acquisition target.

After some near-deals, Schlumberger bought Fairchild in 1979 for $425 million. Schlumberger was a very successful supplier to the oil and gas exploration industry. Over the years I've been really impressed with the way they've done business, but in this instance their past successes led to a terminal case of hubris. They put Tom Roberts, a financial type with no experience as a CEO and no semiconductor background either, in charge. He fared badly. Very, very badly!! The death march had begun! In 1985 Don Brooks, a well-regarded TI executive, replaced Roberts as CEO but the damage had already been done. Revenues continued to decline. Eventually, Schlumberger decided to sell. A Schlumberger spokesman explained, "Silicon Valley ain't the oil business!" No Kidding!! In came an offer from Fujitsu – a huge Japanese conglomerate and a major factor in Japan's plan to dominate the world IC market. The offer was for only $245 million - a small amount for a company with sales of $400 million annually, but Schlumberger jumped at it. Terms were agreed to -- all that remained was government approval. It never came. The US government refused to approve the deal arguing that the sale of a technology company to a foreign entity was not in the best interests of the United States. In the end, Schlumberger agreed to sell Fairchild to National Semiconductor at the shockingly low price of $122 million. To put that in perspective, today TI is valued at more than $100 billion and Intel at more than $200 billion. So ---at its death bed Fairchild was worth about 1/2000 the value of Intel today.

Fairchild started out as the king of the hill. The darling of Wall Street. They made the inven-

---

Note: National spun out a "new" Fairchild in 1997. It wasn't the real Fairchild. They got out of the traditional IC rat race and into new product categories. Power devices. Power discretes. Power analog. High voltage. Opto couplers etc. They were quite successful -- this new "Fairchild" was a winner. But it was Fairchild in name only. It wasn't even close to "our Fairchild." Eventually they were sold to On Semiconductor – a descendant of Motorola. The traditional IC inventor and powerhouse Fairchild was dead. It had died a slow and painful death.

tion that changed the world and enabled a multi-trillion-dollar high tech market which today encompasses computers, cell phones, and all the other amenities that the world takes for granted. But they ended up being virtually worthless.

Why? What happened? In my view there were three major causes.

#1. The exodus

Fairchild could never keep their most important people. Not long after the invention of the integrated circuit, internal strife broke out between some of the traitorous eight. The result was that four of them left in 1961 to become the founders of Amelco. Then, of course, Moore and Noyce left in 1968 to found Intel. The last of the eight to leave was Julius Blank in 1969. You'd think that there would have been great fanfare. There wasn't. One day he was just gone.

The real personnel pirate when I first got there, though, was National Semiconductor. In 1966 Charlie Sporck left his job at Fairchild to head up National. A short while later Charlie recruited a trio of top Fairchild managers including Pierre Lamond. (Pierre eventually became a huge success in the venture capital field. Today he's a partner at Eclipse Ventures. Pierre has a ton of energy!!) Over the next couple of years, many key managers and engineers left Fairchild to go to National.

So ----- Intel wasn't public enemy number one when I got to Fairchild -- National was. But then Intel took their turn at raiding -- and they did an excellent job of it!! Volumes of wonderful engineers and scientists made the jump to Intel. Eventually even AMD took a turn. Jerry Sanders and John Carey, of course, had been fired when Les Hogan came in --- victims of *"Off with their heads."* They went on to found AMD and, I'd imagine, took great delight when their turn came to raid the Fairchild pantry.

The bottom line --- Fairchild just couldn't hang on to their most important employees. The key Fairchild engineers usually did indeed end up making huge contributions. -- They just didn't do it at Fairchild.

#2. MOS happened.

Fairchild started out using bipolar transistor technology. No surprise there. MOS was theoretically a good technology, but in the real world it couldn't be made profitably. The potential benefits of MOS were known , but always just out of reach.

In those days Fairchild didn't really understand mobile ion contamination. Or work functions. Or surface states. Or oxides. To grossly oversimplify, no one knew how to control the thresholds which moved substantially during Op Life tests (We learned about Operating Life / HTRB testing in episode #8. Thresholds shouldn't change during

Op Life tests!) You couldn't reliably turn off N-channel transistors. So -- MOS at Fairchild in the early days was P-channel. And sadly, P-channel MOS was slow! There was one beautiful thing, though. PMOS was a five- mask process. (The most advanced processes today are approaching one hundred masks. Masking steps are the most expensive part of making wafers.) PMOS wafers were really cheap and easy to make so long as you didn't mind bad sort yields and slow, unreliable parts.

CMOS had been conceptualized, but it seemed totally out of reach in those days. It was widely recognized at the Fairchild R&D facility in Palo Alto that there were solutions to these problems and that the upside of MOS (Particularly CMOS) greatly eclipsed that of bipolar. The roadmap was pretty clear: Get rid of the contamination. Make silicon gate work. Switch to CMOS. And finally, scale like crazy!! Scaling helps MOS greatly but helps bipolar only a little. The good news: that road map was followed, and the problems were solved. The bad news: It didn't happen at Fairchild. It happened at Intel. And AMI. And Micron. And Mostek. And many other companies. But Fairchild never really succeeded on the MOS battleground.

Somewhere around 1974, though, Fairchild came up with a counter-punch: The Isoplanar process. A team working under Doug Peltzer developed a new process for making bipolar ICs. The new process used oxide sidewall isolation instead of the traditional reverse-biased junctions. That would make the die a lot smaller and the parasitic capacitance a lot lower if only they could get the yields to a respectable point. After some tough battles with Iceo – the traditional bane of bipolar transistors -- they did it. Ergo --- faster and cheaper!! By quite a bit! Isoplanar bipolar technology staved off the MOS hoards for probably ten years longer than would have been the case otherwise. But the fabs kept scaling. Two microns went to 1.2, then to 1.0, then .8 then .5 etc. etc. MOS kept getting better and better. Bipolar couldn't keep up. Today, for the most part, bipolar is a thing of the past.*

#3. Product planning

In the late 90s we hired a consulting company at Actel. (Name withheld to protect the guilty.) Please shoot me if I ever threaten to do that again! After the normal lengthy and expensive consulting process, the consultants concluded that Actel was too focused on products. I grudgingly accepted that at the time, but I was wrong. The fact was, we weren't focused enough on products. In the non-commodity IC world, your product is all that matters. Branding works well for Apple!! People will go out and buy a product just because it's an Apple product. But, when you're selling to Cisco, what they want is the best product for the

* I was a bipolar transistor engineer. Guess that explains why I'm having such a hard time getting a job.)

job they're trying to do. What does this have to do with Fairchild? Fairchild never put together a product planning system that really worked. Other than Isoplanar bipolar memories and a relatively small line of ECL products, they never seemed to innovate products that customers needed.

Proof of that came when Schlumberger took over Fairchild. They put a ton of capital into the company -- rumors had it that Schlumberger invested the better part of a billion dollars after they bought the company. They bought better fab equipment and better testers. They improved their assembly lines. They also put more money into marketing and selling. But they didn't have a significant product that the world needed. The capital spending was to no avail. Sales didn't rise at all. In fact, they fell. No good products – no sales!

Long story short? Fairchild invented the integrated circuit and kicked off an industry that today hovers at around $400 billion annually. They were at the root of the creation of hundreds of successful companies and many thousands of millionaires. Along the way they helped create probably a few dozen billionaires as well. But, when the dust settled, Fairchild had failed. They were worth next to nothing and the dregs that had some miniscule value lay in the hands of their once most despised competitor.

And Sherman Fairchild turned over in his grave..

Episode 10

Part 2

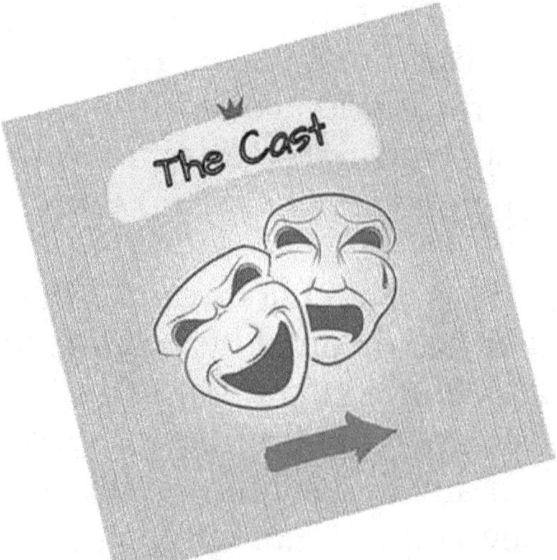
The Cast

The Cast

Charlie Sporck

Image 25

Charlie Sporck ran all of Fairchild's manufacturing operations for seven years. In 1966 he left Fairchild to take the CEO job at National Semiconductor. He took some very key players with him. There was no particular bad blood between Sporck and the remaining Fairchild top-management team, still, a natural Fairchild-National rivalry developed. When I arrived at Fairchild in 1968, National was public enemy number one! (Intel had been formed but hadn't yet appeared in the market.) Soon Fairchild was in a vice. Noyce and Moore led Intel to excellence in the areas of product and process development. Charlie drove National to achieve manufacturing excellence. Fairchild achieved neither. Fairchild was not as good as TI or National in low-cost manufacturing. They weren't as good as Intel at innovation. Over the years those rivalries intensified. If you worked at Fairchild you were expected to hate all three rivals: National, TI, and Intel. That was a lot of hating. Many decided that in truth they hated Fairchild and moved on. A couple of decades later Sporck bought Fairchild from Schlumberger for peanuts -- an ignoble ending for the loyal Fairchild employees who remained. The good news? There weren't many of them. Most had long ago moved on to Silicon Valley start-ups.

Pierre Lamond

Image 26

Pierre Lamond was born in France and received the equivalent of a master's degree from Toulouse University. He took a job in the United States working for Transitron who, at that time, was the second largest producer of semiconductors. In 1961 he was recruited by Gordon Moore to run a development group at Fairchild's R&D facility. 1961 was a bad year for Fairchild. Bob Noyce had invented the integrated circuit in 1959 and a team under Jay Last, after working on it for two years, had finally managed to make a truly functional device. Then much of that team abruptly left Fairchild. There were huge problems!! The specs of the first circuit (Referred to as an RTL circuit) for fan-in, fan-out, propagation delay and noise margin were terrible!!

What does that mean in English? RTL sucked! Those were really the only specs that mattered in those days. The process was poorly defined. The yield was near zero. And there was a reliability problem as well. Pierre took charge of the manufacturing and design group. With the help of a few old friends of mine -- Bill Sievers and Dick Crippen among others -- in a few years he transformed the hideous problem into a very profitable operation. Unfortunately for Fairchild, when Charlie Spork left Fairchild to head up National, Pierre went with him. The constant turnover of Fairchild's best employees was probably the biggest reason for Fairchild's downfall. Pierre later moved into the venture capital field becoming a partner in Sequoia, then Khosla, and finally Eclipse Ventures. Today Pierre and I work together on the board of Tortuga Logic.

Traitorous Eight

Image 27

The traitorous eight, clockwise from Robert Noyce (front and center): Jean Hoerni, Julius Blank, Victor Grinich, Eugene Kleiner, Gordon Moore, Sheldon Roberts and Jay Last.

You met the Traitorous Eight in episode #2. They were the eight brilliant men who left Shockley in order to form Fairchild Semiconductor. (Shockley, of course, coined the term "Traitorous Eight." He viewed them as traitors. That was not even close to the way they saw the situation.). Unfortunately for Fairchild, by the middle of 1968 seven of the eight had left Fairchild -- all to eventually form their own IC start-ups. Only Julius Blank remained when I got to Fairchild in September of 1968. He left shortly thereafter – to form a start-up. Losing that many brilliant people hurt!!!

When the Traitorous Eight left Fairchild, it was the beginning of the start-up rage that Silicon Valley is known for even today. In 1961 Last, Hoerni, Roberts and Kleiner left to found Amelco and then splintered off to form Kleiner-Perkins,* Union Carbide, and Intersil. In 1968 Grinich left to become a professor at Berkeley but then later went on to found Escort Memory Systems and Arkos Designs. In 1968 as well, Noyce and Moore left to found Intel. Finally, in 1969 Julius Blank left and started a business consulting to the dozens and dozens of start-ups who had recently sprung up. Eventually, though, he couldn't stand not being directly involved so he founded Xicor.

---

*Kleiner-Perkins was the first venture capital (VC) company on Sand Hill Road in Menlo Park, California. Today Sand Hill Road is swarming with VCs. It's the center of the world's venture capital business. The prestige of the area has driven the rents to the point that it's the most expensive place in the US to rent office space. Rents on Sand Hill Road are considerably higher than on Fifth Avenue in New York City.

Episode 11

Part 1

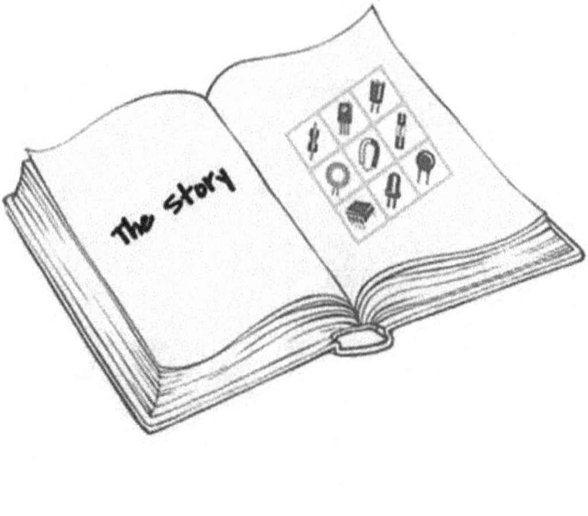
The story

The Story

Real Men Have Fabs

In 1977 I made a job change: I took a job at Raytheon Semiconductor. Raytheon was on Ellis Street next door to the Fairchild "Rusty Bucket." In the early days, they shared the same parking lot so my commute didn't change much, but my outlook on life changed a bunch. I had mostly enjoyed my days at Fairchild, but I hated every single day I spent at Raytheon.

Then, in 1979, I got a break! Gene Conner (a great boss and AMD's first product engineer) offered me a job as product manager of AMD's interface product line. I jumped on it!!! Wow. It was like dying and going to heaven. Within a few days Gene taught me the most important thing that you had to understand if you were going to be a manager at AMD.

People first. Products and profits will follow.

Jerry Sanders was definitely a flamboyant guy. Some of the stories you may have heard are probably overstated, but he was flamboyant! He was also very sensitive to the needs and feelings of the people who worked there. Jerry hated the idea of layoffs. Layoffs are very different from firings.

Someone gets fired if they don't do their job well. It seems harsh, but sometimes that has to happen. With layoffs, though, people who are doing their job well get let go. We all hate that. Jerry particularly hated it. Layoffs were a common part of the Silicon Valley culture at the time (See the episode Layoffs a la Fairchild). Jerry didn't want AMD to be like that. He instituted a no-layoff policy at AMD.

At first it was an informal policy. Later, he had it written in the company's policy manual. For 17 years he stuck to it. If things weren't going well temporarily, Jerry's view was - hold on to the people and let the earnings suffer. Not the other way around. That was unheard of in Silicon Valley semiconductor companies. It made people want to work at AMD.

Then the great semiconductor recession of 1984 came. Things went south fast. We dropped into a loss position. Our spending was too high. Our sales too low. The cash balance wasn't strong. At an executive staff meeting we were hashing out what we could do about it. The subject of a layoff came up. Several execs were pushing for a layoff. Jerry went apoplectic. He banged on the table yelling, "I'm not going to preside over the dismantling of my life's work." Jerry was always a good "quote machine," but that one in particular will stick with me forever.

(Unfortunately, by the time 1986 rolled around we were still in a loss position and the cash balance was running dangerously low. We were forced to abandon the policy.)

In 1980 we had a very good year. Jerry wanted to spread the wealth. He decided to hold a raffle. The winner of the raffle was to get a house! Yes. A real house here in Silicon Valley! Even back in 1980, production workers generally couldn't afford their own houses. The raffle was held, as I recall, on a Saturday night. Early Sunday morning Jerry, accompanied by a Channel 7 TV crew, went to the home of the winner (A Fab worker named Jocelyn Lleno who didn't have any idea that she had won) and knocked on the door. When she answered the door wearing her bathrobe, he told her: "Hi. I'm Jerry Sanders. I came here to tell you that you won the raffle. You've won a house here in Silicon Valley." She was blown away!!! (Actually, the prize was $1000/month for 25 years. Hard to believe, but in those days that was enough to buy a nice house.)

Once at a black-tie dinner event for AMD executives and their wives, I was assigned to sit next to Jerry at dinner. My wife Pam sat directly to his right. Jerry knew that Pam owned a dance studio (she still does). He asked her how the studio was going. It happened that Pam was about to take a contingent of dancers to Russia,

Poland, and the Ukraine for three weeks as part of an exchange program – a cadre of Russian dancers had just visited Silicon Valley. It was expensive to take all those dancers to Russia and nobody had figured out how they were going to pay for it. So, Pam -- extrovert that she is – responded with something like, "Well. I've got a problem. I don't know how I'm going to pay for this Russian exchange. Can you help?" As I crawled out from under the table, I saw Jerry reach into his jacket pocket. He pulled out a check book and wrote out a personal check for $1000.

I first met TJ Rodgers in 1982 when he worked at AMD. Shortly after that, he left AMD to found Cypress Semiconductor. In 1992 plus or minus a year or two, Valerie Rice (a writer for the San Jose Mercury News who told me this story) was interviewing TJ. The fabless concept hadn't yet taken over the world, but it was making inroads. Valerie asked TJ what he thought about the fabless model. I love TJ Rodgers! He was one of the old guard CEOs (As I was). He believed in Fabs, device physics, and transistor level circuit design. (Things have changed. See the upcoming episode: *The Decade That Changed the World*.) Valerie tried to help by summarizing what he had said. "So, you're essentially saying that real men have fabs, right?" That was a play on the title of a book that was very popular back in the day: "Real Men Don't Eat Quiche." TJ jumped on it. "Exactly!!!" Jerry Sanders read that line and loved it! Later that year he was the lunch speaker at the Instat Conference (Jack Beedle's annual semiconductor conference that was attended by virtually all the big brass in the business). The high point of his talk? In his very strongest "take charge of the room and lay down the law" style:

"Now hear me and hear me well. Real Men Have Fabs!!!!"

Most of the speakers that afternoon were fabless company CEOs. I was one of them. Jerry's talk sent us all scurrying back to our power points to make the necessary changes. The Instat Conference was always fun, but that was the best one ever!!

There was something about the AMD environment that spawned CEOs. Was it the collegial environment? In total, at last count 83 former *AMDers* have gone on to become CEOs of other tech companies. The two who impress me the most, though, are two CEOs who were just starting their careers at AMD during the days when I was a VP there. Jayshree Ullal and Jensen Huang. Jayshree (the CEO at Arista Networks) took Arista from a fledgling company to one now valued at twenty billion dollars! There's a great article about her in Forbes Magazine. Jensen (the CEO of Nvidia) has built a juggernaut, but I think of him as the best public speaker I have ever listened to. (Actually – he's tied with Jerry Sanders who I viewed for years as the greatest orator in the history of High Tech!!! They have different styles, but they're both amazingly good speakers!). At the typical dinner event, most of us can't wait until the keynote speaker shuts up so that we can eat. In the case of Jensen, though, you don't want him to stop. He's just plain fun to listen to.

There was a terrific amount of camaraderie and love for the company in the early days of AMD. A terrific spirit! It seemed to me that it waned a bit, though, when Jerry left. On May 8th (of 2019) I was invited to attend the AMD 50th birthday celebration in their new offices in Santa Clara. It was a really well-planned event. I talked briefly with Lisa Su (The new CEO) and Ruth Cotter (The VP of Human Resources). Then I spoke with a dozen or so of the present-day employees. My takeaway? Lisa Su is great. Ruth is as well. The spirit is back.

Jerry Sanders was CEO of AMD for 33 years. TJ Rodgers was CEO of Cypress for 33 years. The industry lost a lot when they retired! I miss them!!

Episode 11

Part 2

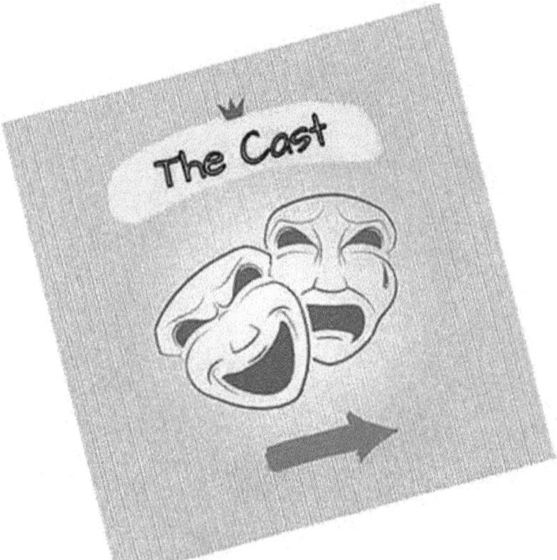
The Cast

The Cast

Jerry Sanders

Image 28

Jerry Sanders was born on the south side of Chicago, the oldest of ten children. He was raised by his grandparents. He attended the University of Illinois on a full scholarship (without that he wouldn't have been able to go there) and received a bachelor's degree in electrical engineering. His first job was as a circuit designer at Douglas Aircraft in Arizona. He moved to Silicon Valley to take a job in sales at Fairchild where he rapidly rose to the top sales job. When Les Hogan and his heroes took over Fairchild, Jerry didn't last long. He was fired in the late fall of 1968. *Off with their heads*. In 1969 Jerry, along with a group of engineers some of whom had also been fired at Fairchild, founded AMD. (The group included John Carey who you met earlier.) Jerry wasn't involved at the very beginning of planning for the new company. The original group approached him and asked if he'd be willing to join them as they needed someone to run sales for them. Jerry responded that he'd be happy to join them so long as they would make him the CEO. They agreed. There were some scary moments in the very early days of AMD. Jerry told me stories about anxiously sorting through the morning mail hoping that there would be a payment from a customer in the mail. If not, there wouldn't be enough money in the bank to make payroll. Then they began to build up a head of steam. In ten years or so AMD eclipsed Fairchild in sales.

AMD is thriving today. Fairchild is gone.

T J Rodgers

Image 29

TJ Rodgers received a PhD in electrical engineering from Stanford University. He took a job at AMI (They sound about the same, but AMI is a different company than AMD) and worked at developing a new process that seemed to have great potential -- V-MOS. V-MOS turned out to have technological hurdles that couldn't be cleared with the technology of the times. TJ moved on to AMD where I met him. At AMD he ran the static RAM product line which had built a very nice business selling N-channel SRAMs. The problem was, by then the world didn't want N-Channel SRAMs. The world wanted CMOS SRAMs and AMD was having big problems getting its CMOS process working. In 1982 TJ founded Cypress Semiconductor and devoted the company in its early days to developing a functioning CMOS process. He succeeded and then rapidly built a substantial business selling CMOS SRAMs. The world had been aware of CMOS and its advantages for many years, but Cypress was one of the first companies to solve the riddle of how to make CMOS with respectable yields. Cypress grew into a multi-billion-dollar company before TJ retired. At the time of his retirement, he had run the company for 33 years. He now devotes most of his time to his winery – Clos de la Tech. He makes an excellent wine --- trust me!

Jayshree Ullal

Image 30

Jayshree Ullal was born in London. She moved to Silicon Valley where she received a master's degree in engineering management from Santa Clara University. She started her career at AMD where she worked in an entry level marketing position in the communications and networking group. When I first met her, she made an immediate impression on me. She was young, but she meant business!! --- She was very serious about getting her job done and done well. From AMD she moved to Ungerman-Bass and then to Crescendo Communications where she was the Vice President of Marketing. Crescendo was eventually bought by Cisco. At Cisco she continued to rise up the ranks until she reported to John Chambers – the

Cisco CEO - running one of Cisco's biggest and most profitable groups. In 2008 she was hired to be the CEO of a hot new start-up -- Arista Networks -- a position she still holds. Arista went public in 2014 and now has a market value of over twenty billion dollars! She was named *one of the top five most influential people in the networking industry today* by Forbes Magazine. Not one of the most influential **women** –- one of the most influential **people**!!

I have two daughters. I love Jayshree's story!!

Jen-Hsun Huang

Image 31

Jen-Hsun "Jensen" Huang, the CEO of Nvidia, was born in Taiwan. His family moved to Oregon when he was still in school. He received a bachelor's degree in electrical engineering from Oregon State and later a master's degree in electrical engineering from Stanford University. His first job was as a junior engineer at AMD. From AMD he moved to LSI Logic and then, in 1993, he founded Nvidia. Jensen realized that the primary job of a processor was moving away from being dominated by processing text and spread-sheet types of numbers. He saw what was coming. The hardest job of a processor was going to be processing and displaying pictures and graphics. Jensen knew that traditional processors weren't very good at that. The rest of the world knew that too, but there was no general agreement on what should be done about it. A better way was needed. At Nvidia, Jensen and his team developed the first commercially successful GPU*. A GPU (Graphics Processing Unit) is a high-speed processing unit that enables real time computer graphics - something which is commonplace today but was exceedingly hard to do at a decent speed before the advent of the GPU.

Nvidia has grown at an amazing pace since Jensen founded it. The Nvidia market cap is now over $300 billion which eclipses the market cap of Intel (About $200 billion) who had been the highest valued US semiconductor company for more than two decades.

*Nvidia's success with the GPU is remarkably parallel to Intel's story with the microprocessor. Neither was the first to come up with the concept. Neither was the first to tackle the problem at hand. But – both were the first to put a good product into the market that actually solved the problem at hand. Both were the first to build successful businesses in their respective markets. Talking about something is easy. Actually doing it is hard. The race goes to the swiftest.

Lisa Su

Image 32

Lisa Su was born in Taiwan. She moved to the United States and received a PhD in electrical engineering from MIT. Her early career was spent at Texas Instruments, IBM and Freescale Semiconductor in highly technical positions of ever-increasing responsibility. She joined a struggling AMD in 2012 as vice president and general manager. Then, in 2014 she was promoted to the position of CEO. During its first 20 years, AMD had flourished. They were a darling of the stock market and a place where everyone wanted to work. I was thrilled when Gene Conner offered me a job at AMD in 1979! By 1999, though, the bloom was off the rose. Tough times had set in. AMD struggled -- maybe even floundered -- for more than a decade. Floundering is not a word that you'd use to describe them now. Lisa has nailed it!! AMD is killing it!! How did she do it? By clearly defining what she wanted AMD to be good at and then executing on that wish. Now AMD is synonymous with high performance computing. It sounds simple, doesn't it? It wasn't. She was named one of the "World's Greatest Leaders" by Fortune Magazine in 2017.

Another great story for women in high tech!!!

Episode 12

Part 1

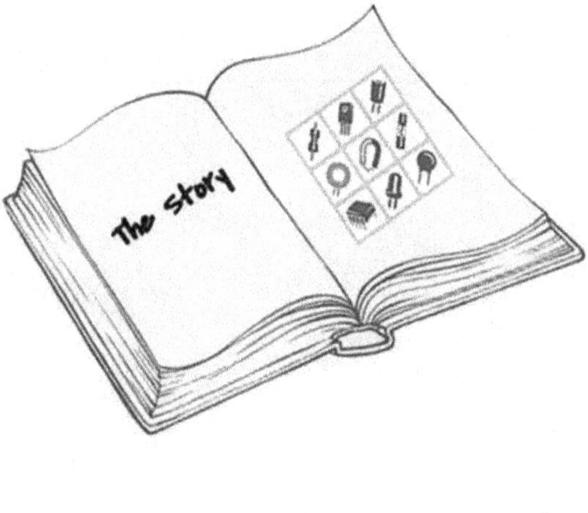
The story

The Story

The Design That Changed the World

Wikipedia ... "In chaos theory, the butterfly effect is the sensitive dependence on initial conditions in which a small change in one state of a non-linear system can result in large differences in a later state". In other words, a butterfly bats its wings in Argentina and the path of an immense tornado in Oklahoma is changed some time later.

In 1980, IBM undertook a very secret project. They had decided to develop a personal computer. Apple Computer was making a killing in the personal computer market. (See the upcoming episodes on Apple and Steve Jobs). IBM owned the big computer market. They weren't about to allow upstart Apple to horn in on their territory! Normal IBM policy was to design their products in a central design group in upstate New York and to use IBM manufactured ICs. They recognized that sticking to this policy would slow things down. They didn't want to go slowly. They wanted to announce the product in the summer of 1981. They formed a task-force group in Boca Raton, Florida working under a lab manager named Don Estridge. The task: get a personal computer on the market and do it by August 1981. Use outside ICs. Use outside software. Do whatever it takes but get it out on time!!! And keep it secret!!!

Meanwhile, Intel was in a tough place. They once owned the memory market (See the episode: *Let There Be RAM*) but now it was becoming extremely competitive. The microprocessor market was becoming so as well. Seemingly every company was offering their own version of a microprocessor. (At AMD we were a microprocessor partner of Zilog who was offering a 16-bit microprocessor called the Z8000.) Over the past decade Intel had gone from a place where they controlled the DRAM and microprocessor markets to a place where they had become just one of the pack. They didn't like that! They created Operation Crush -- a massive project aimed at regaining domination in the microprocessor space. Bill Davidow managed the effort. Andy Grove supported it strongly via a message to the field sales organization saying essentially, "If you value your jobs, you'll produce 8086 design wins." (The 8086 was their newest microprocessor product.)

Paul Indaco (Now the CEO of Amulet Technologies) was a young kid just out of school. He was working at Intel in the Applications Department. As part of a rotational program (common in those days), he was sent out into the field to learn the selling side of the business. As luck would have it, he ended up in the Intel sales office in Fort

Lauderdale, Florida. The custom was (And I'd imagine still is) to give the new guy the account scraps that didn't much matter while the experienced guy kept the important accounts. So --- Earl Whetstone, the existing salesman in the office, took the accounts to the south of Ft Lauderdale and Indaco got the less important ones to the north. One of the accounts that "didn't matter" was IBM Boca Raton. IBM was the biggest high-tech company in the world! How could IBM "not matter"? Because Boca Raton was not a design site. That is, it wasn't where decisions regarding what parts to use were made. Those decisions always came down from Poughkeepsie. --- Or so everyone thought.

One day not long after Indaco had moved to Florida, he happened to be talking with a salesman from his distributor (Arrow). "Oh. By the way. An IBM guy asked me today for some info on the 8086. He works in some secretive new group. He didn't say why he wanted to know." With nothing better to do, Paul got the name and number and called the guy.

Yes. It turned out that IBM was up to something. They wouldn't say what it was. That was top secret. But --- they said they were in a huge hurry trying to make a very short deadline. They said that they had more or less decided to go with a Motorola processor (Probably the 68000) but they conceded that they might be willing to take a quick look at the Intel 8086 along the way.

That wasn't good for Intel. It was generally acknowledged that the Motorola 68000 was technically superior to the 8086. It looked like a longshot for Intel, and they weren't even sure what they were shooting at.

Intel had a few advantages though. The first was their development system -- the 8086 in-circuit emulator. It was better than what Motorola had to offer. That would be helpful in speeding up the design and software debugging process. Given the tight deadline, that could be important! Paul loaned them one. Then came good news. The IBM engineer soon said something like, "Hey, I like this development system, would you loan me another one?" The Intel policy was one loaner to a customer. The issue was clear though: "Any work they do on an Intel development system applies to Intel only, so let's help them do a lot!" So, Paul talked with Arrow who happily agreed to loan three more. IBM often needed help on site from the Intel FAE. (An FAE is a Field Applications Engineer. The customer is the expert of what he/she is trying to accomplish - not an expert of the very complex integrated circuits that they plan to use. The Intels of the world provide expert assistance to their customers. These people are called FAEs.) The project was so secretive, though, that when the FAE went to help with the emulation work, the emulator was separated from the rest of the lab by curtains. All he could see was the door, the emulator, and the curtains. IBM would escort him in, he would solve the problem, and then IBM would escort him out.

Intel had four other advantages: Dave House, Bill Davidow, Paul Otellini, and Andy Grove. Those were good advantages to have!! Dave ran Intel's microprocessor division, Bill ran "Operation Crush," Paul ran Intel's strategic customer accounts, and Andy ran Intel. They wanted this win! Operation Crush was in full force! Any number of issues had to be solved. Among them was the ever-present issue of needing to beat Motorola. And of course, there was the issue of pricing. IBM wanted a price that was in the neighborhood of one half the current 8086 ASP (Average Sales Price). Then, the Intel team had an epiphany! Why not switch from the 8086 to the 8088? The 8088 was an 8-bit external bus version of the 8086. Pricing would be less of an issue with the 8088 and IBM might like it because it would speed up the design cycle. Why? Because Intel had a complete family of 8-bit peripherals which would eliminate the time required to design the functions that the peripherals handled. (Peripheral chips are chips that are used in addition to the microprocessor itself in order to make a complete system) The available peripherals would not only speed up the project, they'd also reduce the number of components required to do the job hence saving money. Saving both time and money?!! Jackpot!! Neither the

68000 nor the 8086 had a complete family of peripheral chips at that time. In the end the Indaco/ Whetstone/Otellini/ Davidow/House/Grove team pulled out a victory. Even after they won, though, they didn't know what they had won until the day IBM announced the product. The design win report that Indaco filed listed a win in a new IBM "Super intelligent terminal." It wasn't. It was the IBM PC.

It ended up being the most important design win in semiconductor history.

What does this have to do with butterflies and chaos theory? ......

Intel is the biggest semiconductor company in the United States. Until recently they were biggest in the world. (Samsung, based in South Korea, recently passed Intel) To a great extent that is due to the IBM design win. I wonder what company would be biggest if Indaco hadn't happened to be talking with the Arrow salesman that day? What if the Arrow guy happened to talk with a Zilog salesperson or an AMD salesperson instead? Or one from National or Motorola or Fairchild?!!! The world might be very, very different!

Grove went on to be Time Magazine's Man of the Year in 1997. House went on to be CEO of Bay Networks and later, chairman of Brocade. Otellini went on to be CEO of Intel for a decade.

Davidow went on to become a very successful venture capitalist with the distinction of leading one of Actel's financing rounds. They all ended up well.

But Indaco has them topped. He went on to become Actel's Vice President of Sales.

Episode 12

Part 2

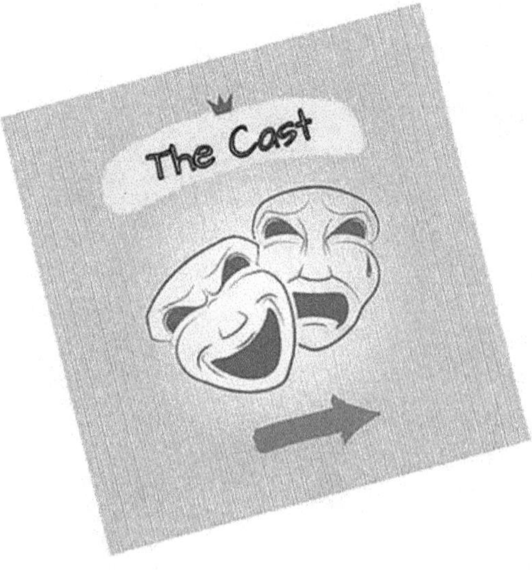
The Cast

with some Intel people regarding technology, patents, and possible patent infringement. Have you ever seen the old movie where an unfortunate cow mistakenly wanders into a piranha infested river? Well -- that was the meeting, and I was the cow. Otellini stepped in, brought some order to the meeting, and saved me. Thanks Paul!

The Cast

Paul Indaco

Image 33

In the IC industry, when you convince a customer to use your product in their new product design, that is called a "design win." By 1980 every IC company knew that the microprocessor market would be big. Nearly every serious IC company including the Japanese companies had either developed their own microprocessor or else had some sort of agreement which allowed them to sell a processor that some other company had developed. Developing a microprocessor was a very tough job. But ... convincing a customer to use your processor proved to be even tougher. There was just too much competition. Exacerbating that problem was the fact that you really didn't know which design wins were going to matter. You might spend an immense amount of time trying to win a design and eventually succeed only to find out that the end-product never sold well. Or – even worse – you might win a design and then be told later that the customer had canceled the project.

On the other hand, you might be totally unaware that a potential customer was working on a project that would turn into a huge success. Later the end-product would be announced using a competitor's product. You would have been shut out of a huge business opportunity due to your ignorance of the situation. Your boss wasn't going to like that!!! You really didn't want to be the salesman in charge of those kinds of situations!

The biggest design win in the history of the IC business was the IBM PC design win. The IBM PC and its later generations have sold in volumes of many billions of units. (Pretty much any machine that runs Windows fits into this category) Winning the first-generation design insured a high degree of success in future generations because of software compatibility issues. Paul Indaco was the salesman handling IBM Boca Raton when Intel bagged the "design win that changed the world." Indaco is seen in this picture holding the award he received later that year for winning the IBM design. They knew that it was an important design win. In retrospect, though, it was far more important than anyone could have imagined at that time. The IBM design win changed the landscape of the industry.

Paul Otellini

Image 34

Paul Otellini was born in San Francisco and earne an MBA from University of California at Berkeley. (Bears!!) He served as the CEO of Intel for a deca but during the time of the IBM design win, he wa a sales-related position -- namely the head of In Strategic Accounts. By the time Paul took the CEO Intel had developed a (well deserved?) reputati having really hard-nosed CEOs and, some woul a culture that could best be described as "Sic e suppose Paul might have fit that role on occasi the Paul Otellini that I knew was a calm, reas levelheaded guy who just wanted to get the rigl with a minimum of carnage being done along In 1990 or so I was running Actel and had a

Andy Grove

Image 35

Andy Grove was born in Budapest, Hungary. When he was 8 years old the Nazi Germans took control of Hungary and began deporting Jews to concentration camps (Andy's family was Jewish). Andy narrowly escaped being sent to a camp. His father did not. At the end of the war, Russia took control of Hungary -- a situation that Grove saw as being nearly as bad as the Nazis. In 1956 Andy escaped Hungary and eventually ended up in the United States where he received a PhD in Chemical Engineering from the University of California at Berkeley. (Go Bears!) Grove met Bob Noyce and Gordon Moore when he worked at Fairchild in the early sixties. When

Moore and Noyce left Fairchild to found Intel, Grove was the first employee that they hired. In 1982 - the year of the IBM design win -- Grove was the COO of Intel under Gordon Moore who was the CEO. In 1988 Grove was promoted to Intel's CEO position.

Andy was not a patient man.

Bill Davidow

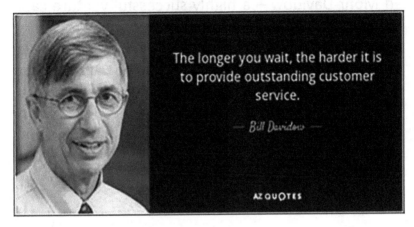

The longer you wait, the harder it is to provide outstanding customer service.

— Bill Davidow —

AZ QUOTES

Image 36

Bill Davidow received a PhD in electrical engineering from Stanford University. He began his career at Hewlett Packard, but eventually moved to Intel where he had a variety of vice-president level jobs. At the time of the IBM PC design win, Bill was running Intel's "Operation Crush". That was fitting since Bill was a key figure in conceiving of, naming, and executing the effort. The concept of Operation Crush was for Intel to recover dominance in the microprocessor market. Largely based on the IBM PC win, that objective was accomplished.

I hadn't met Bill at the time of the IBM design win but developed a close relationship with him when I got to Actel. In fact, he was one of the reasons that I joined Actel – it was clear that I could learn a lot about marketing from him. He did, in fact, have any number of impressive technical degrees, but his true love was marketing. Following his career at Intel, Bill co-founded Mohr-Davidow – a highly successful venture capital company. He is the author of five books: *Marketing High Technology, Total Customer Service, The Virtual Corporation, Overconnected*, and the *Autonomous Revolution*. He now serves simultaneously on advisory boards to Stanford, Cal Tech, and the University of California. He is a very impressive but at the same time very humble man.

David House

Image 37

Dave House received a master's degree in electrical engineering from Northeastern University. After a few years of designing computers for Honeywell and a few other companies, House joined Intel in 1974. By 1978 he had risen to the position of General Manager of the Microelectronics Division -- the division that made the 8088 product that won the big IBM design. During much of the time that House held that position, I held approximately the same position at AMD. As you're going to hear in later episodes, AMD and Intel had a fierce rivalry which included numerous bitter lawsuits based on the rights to the products used in all the generations of IBM and IBM clone PCs. (Basically any computer that ran Windows) We should have hated each other. We might have, except that we didn't meet each other until much later in life. He's a really nice guy. I'm glad it worked out the way it did. While he held that position, House coined the tagline "Intel Inside" which is still used today. He went on to become the CEO of Bay Networks and is now the Chairman of the Board of Brocade Communications Systems. More importantly, though, Dave owns a winery / wine tasting room which is just down the street from where I live – House Family Vineyards.

Life is good!!!

Episode 13

Part 1

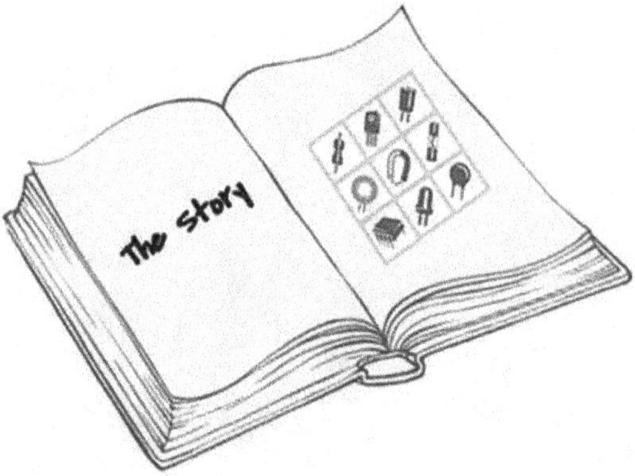
The story

The Story

Steve Jobs, NeXT Computer, and Apple

From time to time I give presentations to various audiences: *Silicon Valley the Way I Saw It.* I always enjoy doing that. One section always makes me stop and think. "Who was this guy? How did he do what he did? Why didn't I do that? What made him so special?" It's the Steve Jobs section. I met twice with Steve Jobs when I was working at AMD. He was a fascinating human being. But the stories about my meetings with him pale compared to the stories about Steve himself. So – let's spend some time going through a little bit of Apple/Steve Jobs history. Apple was formed in 1976 by Steve Jobs and Steve Wozniak. They both went to high school a few miles from where I live, but we didn't cross paths until much later in life. Their first product, the Apple I, was moderately successful – it put them on the personal computer map but didn't set the world on fire. But their second product, the Apple II, did very well -- particularly in the educational market. They went public in 1980. Jobs made two hundred million dollars at the IPO. (An IPO is an Initial Public Offering: In other words – "going public") They kept growing and by 1984 their annual sales were approaching one billion dollars. What a success story!!

In 1983 the Apple board decided that they wanted an experienced CEO. Their view was that selling computers should not be a high-tech sell -- it should be a consumer sell --- more like selling refrigerators than selling mainframe computers. They wanted someone with those skills. They also wanted someone who was more mature than Steve Jobs. Steve was still under 30, had no previous experience at managing anything or anybody before Apple, and was known to be a difficult person to work with. After an extensive search, they found John Sculley, the VP of marketing at Pepsi Cola Company. Jobs loved Sculley. Sculley loved Jobs. Apple really wanted Sculley to join them. At first, he was hesitant, but eventually Jobs convinced him to take the job using the famous line, "Do you want a chance to change the world, or do you want to spend the rest of your life selling sugared water to kids?"

In 1983, of course, DOS based personal computers were a lot closer to mainframe computers than they were to refrigerators. (IBM PCs and their clones used an operating system called DOS which is an acronym for Disc Operating System. It was written and marketed by Bill Gates and Microsoft, but that's another story. Long ago DOS was replaced by Windows.)

To use a DOS computer, you had to talk to it using the DOS language. There was no point and click with the mouse in those days. You had to memorize a cumbersome set of instructions that were part and parcel of DOS. An example of talking to a DOS machine:

DISKCOPY [drive1: [drive2:]] [/1] [/V] [/M]

What the heck did that mean? That was how you went about copying files. Who would want to learn that language? I'd rather learn Greek. Maybe even Swahili. And, of course, the syntax demands were exacting. Any misspelling, extra space or incorrect punctuation would confuse the system. Jobs had an answer for this. The mouse. The point and click, drag and drop interface that he had seen at Xerox PARC. (Xerox PARC was Xerox's think-tank in Palo Alto. The Palo Alto Research Center. They invented any number of wonderful things there including the mouse and some major pieces of the internet. However, the executives at Xerox headquarters were unable to figure out how to capitalize on the wonderful work being done at PARC. Steve Jobs was able!!) That interface – the mouse - was the critical feature of the soon to be announced MAC. Jobs didn't invent point and click, but he recognized its beauty. At his funeral in 2011 his wife Laurene said, "Steve had the ability to see what wasn't there and what was possible." The mouse was a perfect example of that. She also reminded the room of something that Steve had told her. "If a customer is too stupid to use an Apple product, then it's not the customer that's stupid. It's Apple that's stupid!." Again -- the mouse was a perfect example of how Steve thought.

Sculley came aboard, of course, and at first, he and Jobs worked together well. They announced the first Apple MAC in January of 1984 during the Super Bowl. The MAC used a mouse. The mouse made the MAC much, much easier for the layman to use than an IBM or IBM clone. The TV commercial that ran during the Super Bowl, which featured a lady hammer thrower in the days of George Orwell's "Nineteen Eighty-Four," won any number of awards. Even today it's viewed by many as being the best commercial ever made. The relationship between Jobs and Sculley was great! But then it started to fall apart. Soon it got to the point where the board had to choose between Jobs and Sculley. They chose Sculley. Jobs wasn't fired, but he was demoted to a position he wasn't willing to accept. He left shortly afterward in 1985.

When he resigned, he told the Apple board not only that he was leaving, but also, "I'm taking five people with me." The big problem with that? He hadn't told the five people he was going to do that. They hadn't yet agreed to join him. They were afraid that Jobs had told the board who he was taking with him – that might have caused big problems for them. Not the case. Steve didn't tell them who he was taking and amazingly, the board didn't ask. One of the five was Rich Page. Rich was one of four Apple fellows. ("Fellow," is the highest rank that an engineer can attain. Steve Wozniak was also an Apple fellow.) Rich is still active in the Valley today doing angel investing and board work. He (Rich) told me that the five were absolutely shocked when they found out what had happened in that board meeting. But --- Steve really had the power of persuasion! Eventually all five signed up and left with Steve to form NeXT Computer. Steve had a concept in his mind that would lead to what he thought would be the perfect computer for schools. The right feature set. The right price-point. The right introduction dates. It was going to be perfect!!

Laureen Jobs said, "Steve had the ability to see what wasn't there and what was possible." Yes. Usually. But. ----- At NeXT he saw "What wasn't there," but he missed on the "What was possible" part. He envisioned a great

feature set, a great looking box, and a great GUI*
…... But those didn't mesh at all with the great
schedule and the great cost point that he had in
mind.

Steve once said that creativity came from saying
no 1000 times. He did that at NeXT, but it didn't
pay off. The machine came out late and with an
unacceptably high price. Nobody wanted it.

NeXT was failing.

* A GUI, or Graphical User Interface, is the interface between
the computer and the human being using the computer. Before
the advent of the mouse if you wanted to print a document, you'd
have to type out "PRINT" along with some details about the
document you were printing. A graphical user interface allows
you to use the mouse to position the cursor above an icon on
the screen that says "Print". Then you click the button on the
mouse. Voila. The easier your GUI is to understand and use, the
more popular your computer is going to be.

Episode 13

Part 2

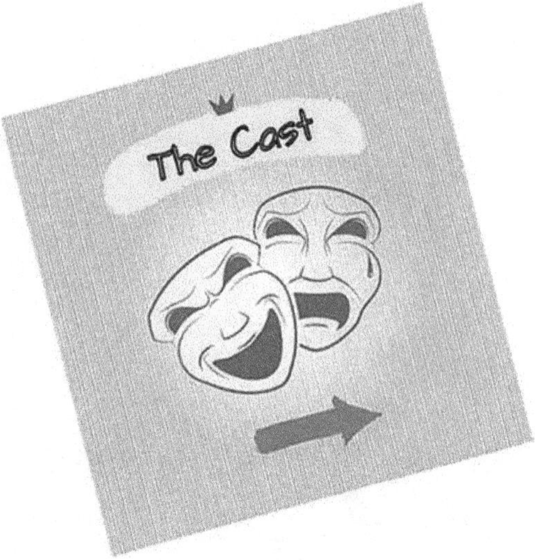
The Cast

The Cast

Steve Jobs

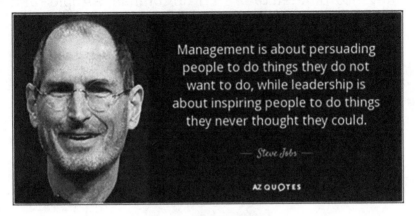

Management is about persuading people to do things they do not want to do, while leadership is about inspiring people to do things they never thought they could.

— Steve Jobs —

AZ QUOTES

Image 38

Steve Jobs was born in San Francisco to an unwed mother and immediately put up for adoption. He was adopted by Paul and Clara Jobs who lived in the Mountain View / Los Altos area in Silicon Valley (Although it wasn't yet called Silicon Valley). Steve attended Reed College in Portland, Oregon for one semester but then dropped out. He didn't want to spend his parents' money on an education that he found to be pointless. His life seemed to wander for a time. He spent several months in India studying Zen Buddhism and several months at a commune farm in Oregon.

Jobs and his friend Steve Wozniak enjoyed tinkering with computers. They attended meetings of the Homebrew Computer Club together. Eventually they decided to start a home-computer company. In 1976 they incorporated their start-up company - Apple - and announced their first product -- the Apple I. Steve Jobs was difficult but brilliant.

Steve Jobs, John Sculley

PHOTO: DIANA WALL

Image 39

Steve Jobs is shown here with John Sculley. You'll hear about my meetings with Jobs in the next episode. I never had the opportunity to meet Sculley. He was brought into Apple as the CEO in 1983.

At first Jobs and Sculley got along well but the relationship soon fell apart and Sculley, along with the board of directors, stripped Jobs of all his responsibilities. Jobs had no interest in being put out to pasture! He immediately left the company. He wasn't a happy camper! Sculley stayed on for ten years but Apple's early successes faded rapidly under Sculley's stewardship. Eventually Sculley got the axe (*Off with their heads* was alive and well in Silicon Valley). Sculley's replacement, Michael Spindler fared no better. He was let go two years later setting the stage for Gil Amelio. You'll soon read more about Gil.

Apple Computer

Image 40

The Apple I computer, Apple's first product, was in truth not an entire computer -- it was just a circuit board. To use it you had to provide your own display (a television set) and your own keyboard. Referencing this picture, the Apple I consisted only of the green PC board in the middle. You had to supply the rest your-self. Apple sold the Apple I for $666.66 --- Wozniak liked repeating numbers. They produced the Apple I for just a little more than one year. They then replaced it with the Apple II -- a big success which really put Apple on the map!! Not many Apple I's were sold and most of those that were sold were destroyed many years ago. The result is that the Apple I is now a rare collector's item. One sold to a museum not long ago for nearly a million dollars. In early 2020, one was listed on eBay at an asking price of $1,750,000.

Steve Wozniak

Image 41

Steve Wozniak was born in San Jose, California. He attended University of California Berkeley (Go Bears) off and on for many years before getting his bachelor's degree in electrical engineering in 1987. In 1971 Wozniak was introduced to Steve Jobs. They formed a friendship that lasted for many years. Together they formed Apple Computer in 1976 and introduced their first product, the Apple I, shortly thereafter. Notwithstanding his education, (he didn't finish college until 10 years later), Wozniak was a brilliant engineer. The Jobs - Wozniak relationship could be summarized as: Wozniak did the engineering - Jobs did the business. That worked well at first but caused problems later.

Episode 14

Part 1

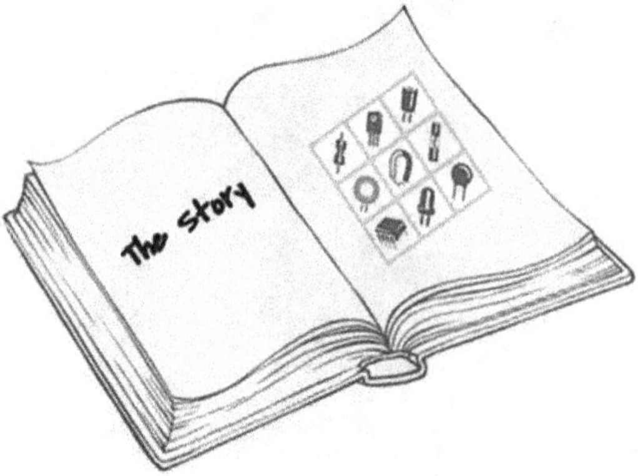
The story

The Story

John Meets Steve

My first meeting with Steve Jobs was in early 1987 when he was running NeXT Computer. I was a VP at AMD and was hunting for potential customers. I visited him in the NeXT Palo Alto facility with the objective of selling him some existing AMD products. He had a different objective: to get me to produce a new product that we had no plans to make but that he felt he needed for his NeXT machine.

Have you ever noticed that dogs can always tell if a person likes them? Somehow, we humans give off some sort of emanation that every dog can pick up. You can't fool a dog. If you like him, he can sense it. If you don't, he can sense it. People are the same in concept, but less sensitive. Sometimes we can pick up the emanation, sometimes not. It depends on how strong the emanations are. Well --- Steve Jobs could really emanate!! You didn't have to be a dog to pick up Steve's emanations. About two seconds after I shook Steve's hand, I sensed --- this guy doesn't like me. He thinks I'm stupid. -- He didn't say it. He was cordial enough in a stand-offish sort of way. But there was no doubt!! But don't feel badly for me. I didn't feel alone. In the course of that meeting, I found out that Apple was stupid too. And John Sculley. And anything to do with IBM/Microsoft. So --- at least I was in good company. But here's my take: he was very, very smart. He'd gone to a liberal arts college for one semester and then dropped out. He'd never had a day of formal training as an engineer. Yet --- I could barely keep up with him as he was describing the technical job that he was trying to get done. He was good.

The second meeting was not that pleasant. I told him that we had decided not to make the part that he wanted. That didn't please him and he said so. He let me know in clear terms that I was not making a smart decision. He also explained that stupid decisions were made by stupid people. QED. He might have been right. In fact, as I look back on it, he probably was right. Rich Page was in those meetings too. Rich was once a fellow at Apple and was now one of the NeXT technical gurus. Rich and I had lunch the other day. We couldn't remember exactly what Steve was asking for, but we could guess well enough to agree that it probably would have been a good product. Oh well. Still --- even though I was never comfortable with Steve, I was in awe of the guy. He was smart, rich, and good looking. When it suited him, he could really turn on the charm! I thought that if he could come up with a little better way of dealing with people that he could own the world. You know what? He did.

Sculley ran Apple until 1993. It wasn't easy! Selling personal computers was nothing like selling mainframes, but it was also nothing like selling soda pop. It was a tough combination of the two. The company languished and Sculley was eventually fired. Mike Spindler took over briefly. Spindler was replaced by Gil Amelio who had been running National Semiconductor. Gil and I worked together at Fairchild back in the 70s. He was a high-level manager when I was just a flunky engineer. I was always really impressed with him. He's a really, really good guy. Smart. Hard working. Good with people. Unfortunately, Steve Jobs didn't see it that way. Why did that matter? Because when he was CEO of Apple, Amelio decided to buy NeXT Computer. When he did, he brought Jobs back as an advisor. No good deed goes unpunished. In a year Gil was out and Steve was back in power.

Steve Jobs was more than just a technologist. In fact, he wasn't really a technologist at all. But, as Laurene Jobs said, Steve could see what wasn't there and what was possible. A great example of this is the iPod. In 2000, Toshiba announced a new, very small form factor hard drive with a 1.8-inch diameter platter -- much smaller than had been available a few years before. A hard drive is used to store large amounts of data. Storing music means storing large amounts of data. Jobs saw the drive at roughly the same time that everybody else did, but his mind put the pieces together better and faster than anyone else. When he put the pieces together, he saw a music player. A year later, in 2001, the iPod was born. Soon it seemed as though every teenager in America had an iPod. I had one too and I was nowhere near to being a teenager. The iPod was a huge win!!

And yet a third example? The MAC, the iPod and then? --- The iPhone!! People wanted to be on the web from wherever they happened to be – not just from their office. And they wanted to do it without having to work hard at it -- that is, they wanted a really simple GUI (Graphical User Interface) – not some impossible to memorize, non-intuitive computer language. And they wanted it on a big, easy to read screen. It wasn't there. But it could be. Apple didn't invent dual touch technology. Apple didn't invent capacitive sensing. Apple didn't invent the ARM 11 processor. (Those three technical advances combined were a huge factor in the success of the iPhone) But Steve saw what wasn't there and what could be. He acted on it. He won again. In the neighborhood of two billion iPhones have been sold.

Over a decade or so Apple introduced iTunes, the iPod, the iPhone, the iPad and finally the iWatch (Apple Watch). All really easy to use. So easy that even a CEO could do it. (My administrative assistant used to use that line all the time) They were good products. I think my daughters are the best barometer in these matters and their houses are full of all of those products and pretty much any other product that begins with an i. What difference did those products make? Apple went from a company who was hemorrhaging cash at the time of Spindler's departure to a trillion-dollar market-cap company. Pretty big difference, wouldn't you say?!!

To repeat Laurene Jobs "Steve had the ability to see what wasn't there and what was possible." How hard could that be? Anybody could do that, right.? Steve got it right almost every time. I tried like crazy to do it but could never quite pull it off. One of us must have been a very special person.

I hope it was him.

Episode 14

Part 2

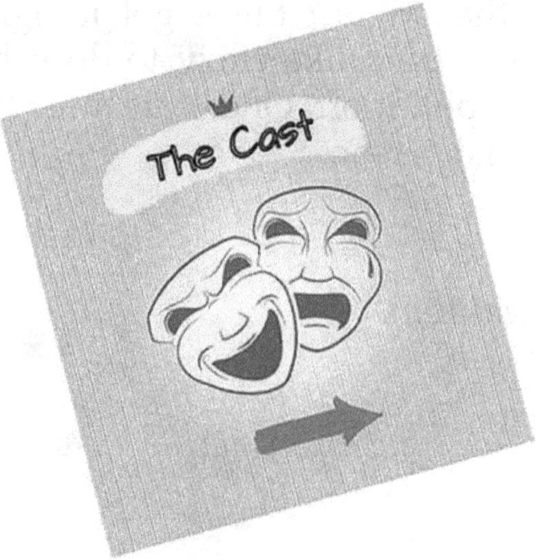
The Cast

The Cast

NeXT Computer

The NeXT co-founders (from left to right) : Dan'l Lewin, Rich Page, Bud Tribble, Steve Jobs, Susan Barnes and George Crow.

Image 42

When Jobs left Apple, his plan was to open a competing company and develop a personal computer aimed at the educational market. Much of Apple's early explosive growth had stemmed from the success of the Apple IIE in schools. Jobs thought that he understood the educational market and also that he knew exactly what kind of product would be a big winner in it. He planned to beat Apple at their own game in the educational market. Nothing would have pleased him more than to look good while making Apple and Sculley look bad by ringing up a big victory there. After all, they had just driven him out of what he still viewed as his own company. (Off with their heads was still alive and well.) This shot shows the founding team of Job's new start-up -- NeXT Computer. The entire team was recruited from Apple. They were set to conquer the educational market. Rich Page is the tall, bearded man standing next to Jobs.

Steve Jobs and NeXT

Image 43

This shot shows Steve Jobs announcing the NeXT computer. He was big on visuals. Every product had to have exactly the right visual effect. He thought that this one did. When asked about that, Bill Gates (The founder of Microsoft and the creator of the Windows operating system) said something like, "Heck. Why not just buy a Windows machine and paint it black?" Two absolutely brilliant people with exactly opposite views on life! Steve, in my mind, was a true genius! It seemed like every market he foresaw came to pass. Every decision he made turned out to be right. Well --- almost. But not this time!!! The product came out late. The planned specs were not met. The cost was way too high. NeXT Computer was a flop. But they had done one thing well. They had developed a good operating system. Apple needed a good operating system. Gil Amelio made the decision to buy NeXT and use the NeXT operating system in future generations of Mac computers.

Rich Page

Image 44

This Pic shows Rich Page and me at lunch. Rich was an engineering fellow at both Apple and NeXT. A huge fraction of Apple's early success stemmed from engineering work that Rich had done there. He was at the top of the list of people who Jobs felt he needed to recruit to join him when he formed NeXT. When I met with Jobs 34 years or so ago, Rich Page was in the meeting. We met for lunch about a year ago. We ate at Rich's favorite restaurant -- Don Giovanni's in Mountain View. Rich and I are both advisors to Silicon Catalyst - a GREAT start-up incubator here in Silicon Valley. He shared some of his best Steve Jobs stories with me. Here's some advice. Invite Rich to have lunch with you. Ask him to tell you a few stories. It will be the best lunch meeting you've ever had!

Gil Amelio

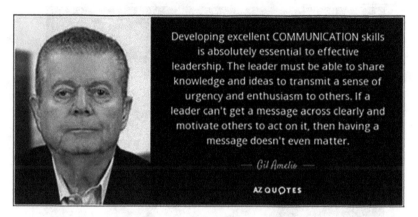

Image 45
Image from azquotes.com licensed under CC BY-SA 4.0

Gil Amelio was born in Miami, Florida. He received his PhD in physics from Georgia Tech. He was a principal member of the team at Bell Labs that developed charge coupled devices -- devices which were, for many years, the technology behind digital cameras. Gil was on a managerial fast track. He ran the CCD group at Fairchild Semiconductor where I first met him. He moved on to be the president of Rockwell International's semiconductor division and then to be the CEO of National Semiconductor picking up where Charlie Sporck left off. (You met Charlie in earlier episodes). After all that success, Gil accepted the job of CEO of Apple. It didn't work out well for him. After a little more than one year in that job, Apple let him go.

Episode 15

Part 1

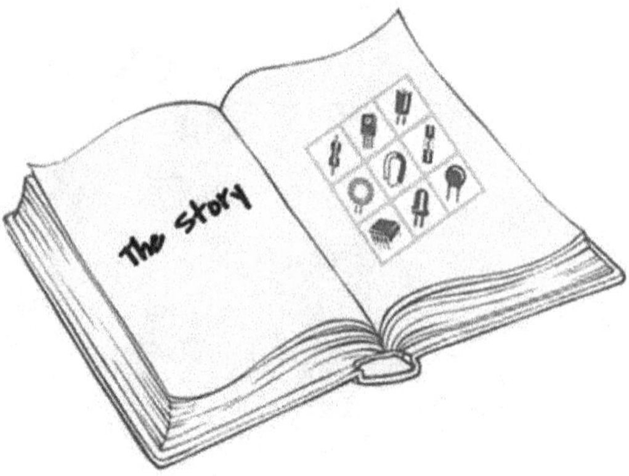
The story

What Fairchild and Jurassic Park Have in Common

Me. I'm a Dinosaur!

In the early episodes of this book, I talked about what I believe were the three seminal events in the history of the semiconductor: Shockley's invention of the transistor, Noyce's invention of the integrated circuit, and Intel's 1971 -- the introductions of the first commercially successful DRAM, EPROM, and microprocessor. I was taking some poetic license when I talked about Intel because that wasn't really an "event." It was a series of events that took place over a year or two. (It should be said here that, as is true in pretty much all inventions and product introductions, if the whole story is told, scores of people and companies would join in getting credit.)

Given that I've already gotten away with a loose definition of "event." I'm now going to talk about a fourth seminal "event:" A decade long event --- one that changed the industry and left us old guys scratching our heads wondering what had just happened.

When I first joined AMD, the marching orders were: "Building blocks of ever-increasing complexity!!" It made sense. For most of the '60s the IC market was composed of TTL "small scale integration" products --- gates and flip flops. (See episode #8 of this book. *Texas Instruments and the TTL Wars*) We called those "SSI". In the late '60s "Medium Scale Integration" (MSI) emerged. Companies that bought gates and flip flops invariably used them to make a generally accepted array of larger elements: multiplexors, decoders, register files, adders, shift registers, counters, and ALUs were on the top of that list. Those functions became the industry's standard MSI products. Every company who was designing electronic equipment would use products out of that group. So - as we progressed along the Moore's law curve, we knew what to do with the additional gates. Instead of selling gates and flip flops, we sold decoders and multiplexors. More progress along the density curve led to "Large Scale Integration" (LSI). AMD's building block concept called for defining generic blocks that could be used by a variety of different customers in a variety of different applications. When AMD got to the Large-Scale Integration point on the Moore's Law curve, they took their best shot with the

AMD 2901. The 2901 was a four-bit wide slice of a processor's Arithmetic-Logic Unit. (What's that in English? Put sixteen of them together and you get the portion of a modern processor that performs calculations.) It hit the bull's eye! Thanks John Mick. Thanks Tom Wong! Thanks John Springer!!

The 2901 was a big success commercially. Just as was true of the MSI building blocks, you could pretty much use the 2901 in any kind of electronic system. But would it always be that easy? Were there plenty of products like the 2901 at hand? And --- looking ahead --- was there a commonly accepted set of VLSI (Very large-scale integration) products that could be used anywhere when we got to that point on the curve? Not really. There were a few products, but it was a much shorter list. Microprocessors and memories could be used everywhere. They were VLSI. Gate arrays could as well (Gate arrays are explained in the "cast" of episode 16) --- but gate arrays were custom products. What else? Not much. "Building blocks of ever-increasing complexity" was reaching the end of its rope. As time passed and IC densities increased, customers ceased wanting to use building blocks to build their own systems. When densities got to the point that the entire system could be put on a single chip, then that's what the customers wanted. Why go through the hassle of designing the system when you could just go out and buy it?

So --- the game had changed. From the IC house's point of view, instead of designing general purpose building blocks that could be used for pretty much any system, IC houses had to pick a particular system and then design chips specifically for that purpose. One chip for switches. One for ISDN. One for ethernet. One for FDDI. Etc. That was problematic on two fronts. First, which systems would you pick? If you picked FDDI and ISDN (As I did) and those markets never went anywhere (As those two didn't), then you'd eat the development costs, and your sales would languish because you wouldn't have the products that your customers wanted. Second, how would you learn enough about the specialized market you were going after to be able to define and design a good product? After all networking experts, for example, by and large preferred working at networking companies – not at chip companies. (ISDN? FDDI? What do all those four-letter acronyms mean? What do the letters stand for? *Who cares? Doesn't matter!* Save your brain cells for more valuable knowledge. Those markets and products all disappeared taking my R&D dollars with them.)

From the customers' point of view, by the time 1988 rolled around they were demanding microprocessors, memories, and gate arrays. Customers could use gate arrays to design their own LSI chips. They liked that!! And virtually all systems use memories and microprocessors. The Japanese saw this coming. They coined the term "master slice" (the term they gave to gate arrays). Then they embarked on an effort to conquer the market for the three M's: Memories, Microprocessors, and Master Slices (gate arrays). It seemed for a time that they might succeed. Basically, their plan was to dominate the business of manufacturing ICs. The good news is that they didn't succeed. (The bad news is that Taiwan and South Korea eventually did.)

Coincident with this was the advent of the foundry. Morris Chang saw the future!! TSMC happened! TSMC took on the responsibility of handling the processing and manufacturing issues that were the bane of the early Silicon Valley crowd. Semiconductor companies no longer had to understand Iceo, or work functions, or mobile ions, or minority carrier lifetimes or any of the many other time-honored problems that had existed since the days that I was a hands-on engineer. Those problems would be handled by some man or woman in Taiwan whom you had never met and would never need to meet. The world had changed. In 1980 the typical successful semiconductor CEO understood semiconductor physics, fab processing, and transistor level circuit design. By 1990 that set of knowledge was already rendered near-ly useless. By 1990 you needed to understand the architectures of very specialized, complex systems and the end markets of these systems. Were there exceptions besides the three M's? Standard products that could be sold to nearly everyone? A few. Notably some analog functions and programmable logic. But not many!

Today I'm spending quite a bit of my time with semiconductor start-ups. Over the past four years I've been working with Silicon Catalyst. Silicon Catalyst is an incubator who incubates only semiconductor companies. That has allowed me to get a good look at dozens of start-ups in the field. My take? Successful semiconductor CEOs today often know little or nothing about semiconductors per se. Moreover, they don't even care!! Nor should they!!! They know about the market they're trying to serve. They understand the hardware, firmware, software and applications related to their chosen market. Building blocks of ever-increasing complexity went out the window a long time ago. Today the call is for a complete solution to an existing problem. Let TSMC worry about how to make the things. The change in the attitudes of customers caused a corresponding change in the strategies of the traditional IC companies. In the case of AMD, "Building blocks of ever-increasing complexity" morphed into "High performance computing is transforming our lives." It was a big change. Companies that don't embrace change wither away. In the case of AMD, they made the changes well. The stock market now pegs their value at north of $30 billion!!!

(For a better understanding of the fabless / TSMC revolution, read Daniel Nenni's Fabless: The Transformation of the Semiconductor Industry. Dan is the CEO of SemiWiki.)

Episode 15

Part 2

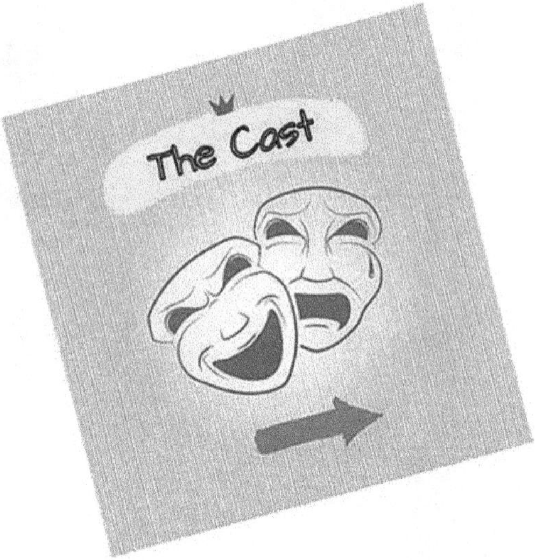
The Cast

The Cast

TSMC/Morris Chang

Image 46

Morris Chang was born in China and moved to the United States when he was in his teens. He attended Harvard and MIT before eventually getting a PhD from Stanford. His first job was to develop a line of transistors at Sylvania Semiconductor, but most of his career prior to TSMC was spent at Texas Instruments where he eventually ran TI's entire semiconductor business. TI was by far the largest manufacturer of semiconductor products at that time. Morris was the person who put TI's forward pricing model in place. The forward pricing model involved quoting what seemed like ridiculously low prices for distant future deliveries of their products. He saw that, because of the progress being made, a low price today would be a reasonable price tomorrow so TI would, in fact, be able to be profitable at those prices.

The benefit came when competitors saw the low price and decided to stay away from that market because of the crazy, seemingly unprofitable pricing. The strategy worked for a while but eventually many, many companies entered the TTL market. Still, TI was able to maintain dominance in the market.

From TI, Chang moved on to become CEO of General Instruments for two years. Then he returned to Taiwan. The start-up craze was still alive and well in Silicon Valley. Morris saw that there were countless companies with ideas for nice semiconductor products – products with the potential for sales of, maybe, 50 or 100 million dollars a year. He also saw what was happening to the cost of a top-notch wafer fab – the area where ICs were manufactured. It was headed north. The cost was soon to pass through a billion dollars on its way to ten billion dollars – roughly where it sits today. A young company with a $50 million per year product couldn't afford a fab to make that product in. Morris saw the potential for a semiconductor foundry - a huge fab that served the needs of all the smallish semiconductor companies. TSMC was born. TSMC (Taiwan Semiconductor Manufacturing Company) is now a $50 billion per year company who serves IC companies all over the world.

Wafer Fab

Image 47

A wafer fab (The area where integrated circuits are made) could be built for a few hundred thousand dollars when I started in the business. Today a good one costs in the neighborhood of ten billion dollars and that cost is going up more every day. In the early days every semiconductor company had their own wafer fab. The capabilities of their fab determined, to a large extent, the success of the company.

Today, though, only the very largest semiconductor companies can afford their own wafer fab (Intel, Samsung, and Micron for example). All the rest use semiconductor foundries -- third party wafer fabs that offer manufacturing services to the rank and file of semiconductor companies. Morris Chang saw this sea-change coming, and started TSMC – Taiwan Semiconductor Manufacturing Company. TSMC soon grew into a huge company with tremendous influence in the semiconductor market.

In the early days, the number one worry of a semiconductor CEO was his or her fab and the processes that ran in the fab. At times that was an all-consuming set of worries because manufacturing a semiconductor is an extremely complicated process. Once TSMC relieved semiconductor companies of those worries, the CEOs were free to spend their time worrying about products, customers, and markets -- the things that really should matter. Today a semiconductor company designs a product and sends a mask-set (Roughly speaking a set of blueprints) to the foundry and the foundry takes it from there. It sounds great, but there is a problem -- the center of excellence for integrated circuit manufacturing has moved from Silicon Valley (as well as Texas where Texas Instruments had their fabs and Vermont where IBM had theirs) to Taiwan. Given the importance of semiconductors in our everyday life as well as in the defense of our nation, the situation today can be a bit troubling.

Episode 16

Part 1

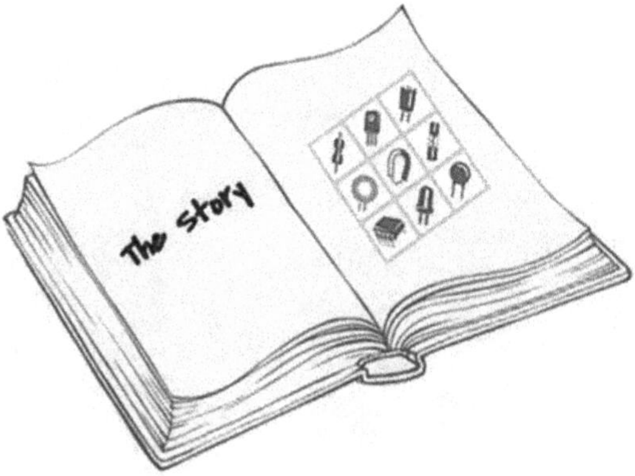
The story

From AMD to Actel

By the late 80s it had become clear to me that the Japanese were right. Memories, Microprocessors, and Gate Arrays (Gate arrays are explained in the "cast" section) were what customers wanted then. "Building blocks of ever-increasing complexity" was obsolete. What next? Should I try to become an overnight networking expert? Maybe a DSP expert? Pretty tough to do. Or --- how about programmable logic? That could work!!

MMI (Monolithic Memories Incorporated), using technology developed by John Birkner and HT Chua, had introduced successful field programmable logic products a decade earlier using nichrome fuses – devices that conducted current unless you forced a very high current through them and "blew them out." The high current destroyed the fuses so that they would no longer conduct current. This "off" state would be the new permanent state of the fuse. This was exactly the same concept as the fuses that were in the fuse box of every home back when I was young. (Remember – *I'm a dinosaur.*) MMI's programmable logic was great. People loved it. But -- the programmable circuits available then (Called PALs) were small --- on the order of a few hundred gates

when the real demand was for circuits with capacities of many thousands of gates. It was clear
to me that a truly field programmable gate array
would be a big winner.

I was approached by the board of directors of Actel. They were looking for a new CEO. Their product was to be a field programmable gate array.
(Andy Haines came up with the term "FPGA"
at Actel in 1989 so they weren't using the term
"FPGA" when they were trying to hire me, but
that's what it was.)

By then programmable logic had advanced sig-
nificantly since the MMI/PAL innovation. Under
the leadership of Bernie Vonderschmitt, a start-up
named Xilinx had introduced a new concept which
used large numbers of flip flops instead of fuses
to store the configuration data. Flip flops were
standard elements which were made on the standard processes* available from the foundries ---
there was no need for Xilinx to try to develop a
custom process. (semiconductor "processes" are
explained at the end of this episode) Although I
failed to recognize it at the time, this would turn
out to be a huge benefit!!! (I had been thinking
that it was a modest benefit. Wrong. It was huge!!)
Xilinx referred to these products as "SRAM Logic Cell Arrays." Later they changed that to SRAM
FPGAs. Actel came along five years after Xilinx
with a plan to use antifuses instead of flip-flops.

An antifuse is the opposite of a fuse. Antifuses do not conduct electricity unless a very high voltage is placed across them. When that happens, the fuses are "burned in." In other words, they will now conduct current. This "on" state is the new, permanent state of the antifuse. Antifuses require a custom process.* Custom processes spell trouble!!! I figured that out later. ☹

It seemed to me that SRAM FPGAs had a glaring defect. There are many ways that the state of a flip flop can accidentally be changed. A power supply spike. A power supply brown out. A hit by a proton or ion or by some stray radiation. These are all unpredictable ways that configuration can be lost. To make the problem worse, there was never a good way to know in real time that corruption had occurred. (I think that this is true even today). So -- if corruption occurred, the circuit might go merrily on its way producing incorrect results and the user wouldn't know. It was a recipe for disaster!! Or so I thought. As mentioned above, the SRAM technology did have a significant advantage --- it allowed you to use a standard foundry process.* The antifuse required a handful of tricky extra process steps the foundries didn't like. But --- overall I felt that the antifuse would prevail. The antifuse process would eventually become a standard and the reliability benefits accruing to the antifuse would win the day!! Again --- so I thought.

221

Actel's board of directors looked like an all-star team. Ed Zschau (formerly in the US House of Representatives) was on it. Bill Davidow (One of the most famous marketing guys in the IC industry) was on it. And Carver Mead was on it as well. Carver was a Cal Tech professor with an IQ of about 200 who had co- written a textbook that was used in virtually every university. (The textbook was known as "Mead-Conway"). After some foreplay, Ed Zschau invited me to have breakfast at his home in Los Altos. During the meal, he offered me a job as the CEO of Actel. Wow!!! I had an opportunity to become the CEO of the only antifuse based FPGA company. I love FPGAs! Antifuses will dominate the industry!!!! I'LL TAKE IT!!! WHEN DO I START? Boy. Was I ever wrong!!! Sometimes I feel like I've spent the better part of my life being wrong, but that was the pinnacle. The antifuse did not turn out to be a wonderful thing! Why? First, customers wanted reprogramability -- the ability to change the configuration as many times as might be needed without having to remove the part from the board and pay for a new part. Antifuses were OTP (One time programmable). If you wanted to change something in an antifuse FPGA, you were out of luck. You had to unsolder it from the board, throw it out, buy a new one, program it, and solder it back onto the board. Customers *hated* that!!! Second, antifuses required a difficult custom fab process.* That prohibited us from ever being on an advanced process node. (Xilinx was always on the most advanced process nodes). Those two problems prevented us from competing well in the two biggest and fastest growing market segments – telecom and networking.

I started at Actel in 1988. By the mid-90s my love affair with antifuses had all but ended. It had become clear to me that the SRAM configuration technique was clearly better for the majority of applications in those days. Xilinx and Altera were already there with SRAM products. We had only antifuse products. There's an old business school case study that goes something like this: Company A dominates a market. Company B comes along later with a better way to do it. Who wins? It's not obvious. It should shape up into a good fight. That's the battle I thought we'd be fighting when I joined Actel. It wasn't. The battle we actually fought was: Company A dominates a market. Company B comes along later with a worse way to do it. Who wins? Duh. ….. By the mid-90s I could see that we would eventually die fighting that battle unless we changed the rules of the game. Clearly, we had to replace our antifuse technology with something better. With what? Good question!! I wasn't sure. What was clear, though, was that it was going to take time and money! It wouldn't be fast. It wouldn't be cheap. And it wouldn't be easy!! Meanwhile, the job was to figure out some market segments where we could at least compete well in the interim.

My favorite marketing book is the 22 Immutable Laws of Marketing. It's only 140 pages, but they're 140 pages of real wisdom. Chapter 2 tells you in essence, "Don't try to horn in on a market that someone else already owns. Figure out a subset of the market that *you can own* and go after it!!!" And we did. Our successful subsets generally centered around the reliability issues that I spelled out earlier. The big markets like communications and networking weren't buying into the antifuse concept. But -- markets that were super-worried about reliability often saw the Actel advantages. The best of these markets for us turned out to be the satellite market. In general, though, we did pretty well in the military and aviation markets as well as with certain industrial and medical products. Those markets were good to us. They allowed us to be consistently profitable. But ---- they were all slow growing markets. Wall Street wants fast growing markets and fast-growing sales. I always wanted to be growing faster, but it wasn't to be. Life wasn't easy in the antifuse business.

Ten years ago, Microsemi bought Actel. Three years ago, Microchip bought Microsemi. Today, the old Actel is a division of Microchip – a highly successful company based in Phoenix.

So ---- what became of the antifuse? Actel still ships some antifuse products --- mostly into designs that were won long ago --- but it's no longer an antifuse technology-based FPGA company. Under the leadership of Esmat Hamdy, Esam El Ashmawi and Bruce Weyer, (Esmat is a founder who directs technology development. Esam and Bruce took turns running the company after I left), Actel completed the long strived-for transition into being a flash-based FPGA company. (To be exact, flash and other newer reprogrammable non-volatile technologies. "Non-volatile" is explained in episode 6.). Flash FPGAs retain the benefits of antifuses but are reprogrammable (they can be erased and programmed differently many times) and use a process that's much closer to standard. Steve Sanghi made a wise acquisition! ** Actel and Microchip are thriving together thanks to the folks who have been there making things happen for so long - many of them for 20 years or more.

So --- Thanks Esmat, Thanks John. Thanks Ted. Thanks Lisa. Thanks Jon. Thanks Sinan. Thanks Ken. Thanks Fei. Thanks Joel. Thanks Arun. Thanks Jerome. Thanks, Toufik. Thanks Nizar. Thanks Frank. Thanks JJ. Thanks, Fethi. Thanks Salim. Thanks Greg. Thanks, Sifuei. Thanks Raymond. Thanks, Habtom. Thanks Antony. Thanks Kathleen. Thanks Melissa. Thanks, Nuru. Thanks Justin.

Thanks Hui. Thanks Cathy. Thanks, Chih Ping. Thanks, Alynn. Thanks Rick. Thanks Paul. Thanks Gina. Thanks Pam. Thanks Vidya. Thanks Ken. Thanks Val. Thanks Volker. Thanks Dirk. Thanks Amir. Thanks Gina. Thanks Jose. Thanks Max. Thanks Trina. Thanks Dora. Thanks Norma. Thanks Lisa. Thanks Carlos. Thanks Mateo. Thanks Hung. Thanks Dave. Thanks Bruce. Thanks David. Thanks Ray. Thanks Frank. Thanks Nelson. Thanks Becky. Thanks, Vinh. And thanks Tessie!!

We owe you a lot!!

What's a "Process"? Making an IC is very complex. It involves many hundreds of steps. The steps performed coupled with the sequence in which they are performed combine to make what is referred to as a "process." It takes a lot of effort and costs a lot of money to design a process, debug it, and get it into production. Once that work is done, a foundry naturally will want to use that process – not to immediately develop a new one.

** *Steve Sanghi is the CEO of Microchip.*

Episode 16

Part 2

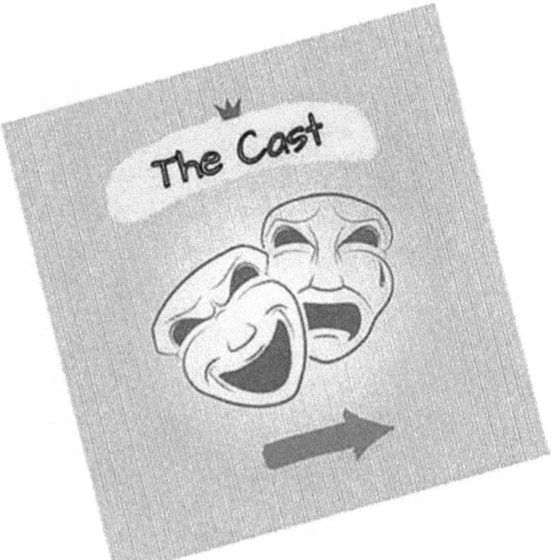
The Cast

The Cast

Gate Arrays, FPGA's and ASIC's

Image 48

Gate Arrays, FPGAs, and ASICs? (Oversimplifying greatly as usual). There are three fundamental types of digital circuits. In episode #6 you learned about processors and memories. The third type is harder to explain. In the most general sense, the third category comprises anything that doesn't fit neatly into either of the first two. Let's call it "logic." Memories and processors are standard products. Two totally different systems may use the same exact processors and memories. Those systems are differentiated by the "logic" that they use. Every system is different, so in the early days the "logic" was implemented using the SSI and MSI circuits discussed in episodes #8 and #11. LSI Logic (and many other companies) developed a relatively fast and simple technology for putting large numbers of SSI and MSI circuits on the same chip and wiring them together to make a custom chip. The products produced using this technology were called "gate arrays." You heard a little about LSI Logic and Wilf Corrigan in episode #9.

Since gate arrays are custom products, no one except the company who designed and ordered them can have any use for them. So --- they carry a burden. Inventory logistics. If you don't order enough, you may run short and need to reorder. That can typically take a couple of months before you can get the material you need to build your system and ship to your end customer. Sometimes longer! That's worse than it sounds. Your end customers will not want to wait an extra two months to get the product that you promised them. On the other hand, if you ordered too many you'll end up having to eat them – to pay for them and then throw them away. No one else will have any use for your custom parts.

Actel, Xilinx, and Altera independently developed techniques for customizing gate arrays after they have been shipped to the customer. Those products are called FPGAs. Field Programmable Gate Arrays. An FPGA is a standard product until your customer "customizes" it just before shipping the product to the end user. Before that, an FPGA is a standard product. It can be bought from a distributor who keeps a stock of FPGAs and who can ship from that stock on very short notice. If you order too many, you can return them to the distributor who can ship them to someone else the next day.

Over the last decade or two, gate arrays have been mostly displaced by even newer technologies referred to as "standard cells" and "ASICs" (Application Specific Integrated Circuits). In episode #6 you learned about the difference between volatile and non-volatile memories. Similarly, there are volatile and non-volatile FPGAs. You heard a bit about that in this episode. SRAM FPGAs are volatile. Antifuse and FLASH FPGAs are non-volatile. More detailed discussions of these topics are beyond the scope of this book.

All three of the original FPGA suppliers eventually merged into larger companies. Actel is now part of Microchip Technology. Altera is now part of Intel. Xilinx is soon to be part of AMD.

John Birkner and HT Chua

Image 49

John Birkner and HT Chua were working together at Monolithic Memories (MMI) when they came up with an idea -- a new way to implement programmable logic. The concept of programmable logic had been around for a long time. The idea was to put switches into an IC in such a way as to allow the user to "program" the chip long after it had been shipped out of the factory and just before the customer was ready to use it. In concept you could program the chip to do anything you'd like it to do. In other words, each customer could build his/her own custom chip. Many companies had dabbled with the concept of programmable logic, but MMI was the first to strike gold -- to introduce a product to the market that was generally accepted and sold well. John got an engineering degree from UC Berkeley. (Go Bears) By coincidence he lived next door to me in the dorm my freshman year. HT Chua was born in Malaysia. He moved to the United States and received a master's degree in electrical engineering also from the University of California at Berkeley (Go Bears). After leaving MMI, HT and John went on to found Quicklogic who, no surprise, based their business plan on programmable logic.

Ed Zschau

Image 50

Ed Zschau was born in Omaha Nebraska. After studying at Princeton, he moved to the West Coast and received a PhD from Stanford University. In his youth Zschau founded System Industries and was its CEO until he ran for congress. From 1983 to 1987 Ed served in the US House of Representatives as the representative from California's 12th district. Upon retiring from political life, Ed spent several years in the venture capital business as well as serving as the CEO of Censtor and Adstar at various times. After retiring from the high-tech rat race, he became a professor at Harvard and a visiting professor at Princeton. You'll never meet a more impressive (Or a nicer) guy than Ed.

When he offers you a job, you'll take it!! I did.

Bill Davidow

Image 51

We met Bill Davidow earlier in the IBM design-win episode. After leaving Intel to found Mohr-Davidow Ventures, he decided to make an investment in Actel which, at the time, was an early stage start-up that was still operating from hand to mouth on a small amount of seed funding. After investing in Actel, he served on Actel's board of directors for several years. He's been very successful both in his operating roles at Intel as well as in his venture capitalist endeavors.

Still, I'm most impressed by the educational and literary aspects of his life. A bachelor's degree from Dartmouth. A master's degree from Dartmouth. Another master's degree from Cal Tech. And a PhD from Stanford. To date he's written five books. Good Books!!! Full of real wisdom!! The most recent -- the Autonomous Revolution -- was recently published. I haven't read it yet. It's sure to be good! I'll start it as soon as I finish writing *Silicon Valley the Way I Saw It*.

Carver Mead

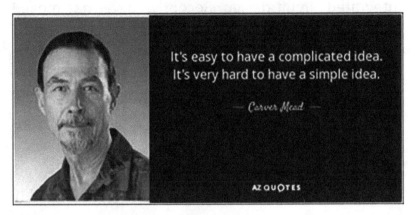

It's easy to have a complicated idea.
It's very hard to have a simple idea.

— Carver Mead —

AZ QUOTES

Image 52

Carver Mead was born in Bakersfield, California. He received his PhD in electrical engineering from Cal Tech. For the next 40 years he was a professor of electrical engineering at Cal Tech. In the sixties, Mead worked closely with the Fairchild founders -- particularly with Gordon Moore. In fact, Mead was the one who gave the name "Moore's Law" to Gordon's prediction that the number of transistors in a chip would double every year or so. It was in the early seventies that Mead began to see the future -- constant scaling of the transistor to ever decreasing sizes. He recognized, contrary to conventional wisdom at the time, that everything got better when you made a transistor smaller -- speed, power, cost --- everything!! He was the first to envision the benefits of sub-micron scaling -- the very reason we've been able to make the huge progress that's been made. In 1979 he and Lynn Conway teamed up to write *"An Introduction to VLSI Design"* -- a text that was used by dozens of universities in their courses on integrated circuit design. Recently Carver has moved on to study general relativity. He believes that Einstein may have gotten it a little bit wrong when he developed his theory of gravity and general relativity. Knowing Carver, I suspect that he may be right. Needless to say --- don't try to take him on in an IQ contest!

Bernie Vonderschmitt

Image 53

Bernie Vonderschmitt was born in Jasper, Indiana. He received a master's degree in electrical engineering from the University of Pennsylvania. In his youth Bernie worked at RCA and, in fact, was one of the key figures involved in developing color TV. (Yes. As strange as it sounds now, there was a time when television was only available in black and white). In 1984 Bernie teamed up with Bill Carter and Ross Freeman to found

Xilinx. Bernie was the CEO of Xilinx for more than a decade. Xilinx was the first company to develop and market what today is called an FPGA (Field Programmable Gate Array) although they didn't call it by that name until much later. I had every reason to dislike Bernie -- we were involved in a bitter patent dispute. But --- I couldn't do it. Bernie was a world class human being!!! There were once nearly 60 competitors in the FPGA market. Gradually almost all of them either exited the business or were acquired. The big three during my tenure - Xilinx, Altera, and Actel -- were all acquired.* Today the largest remaining independent supplier of FPGAs is Lattice Semiconductor. Lattice started as a major factor in the PAL business and gradually morphed into a successful FPGA supplier.

*The AMD acquisition of Xilinx is due to be completed by the end of 2021.

Steve Sanghi

Image 54

Steve Sanghi was born in India. He moved to the United States and received a master's degree in electrical engineering from the University of Massachusetts. After a long stint at Intel and a short one at WaferScale Integration, Steve took the job of president of Microchip Technology. Shortly after that he was promoted to CEO. When Steve took over, Microchip was losing money and running out of cash. It was a scary time for them. Steve drove sales up and expenses down. Just three years later, in 1993, things had improved enough that they were able to execute an Initial Public Offering. (also known as an IPO or "going public.") It was a very successful offering. Fortune Magazine named it the best performing IPO of the year. I was happy for

Steve notwithstanding the fact that Actel went public in 1993 as well. I wanted the award. Steve got it. But only because he deserved it. Who says life isn't fair? Microchip has continued to grow at a rapid rate. Now their annual sales exceed $5 billion. Three years ago, Steve bought Actel as part of Microchip's acquisition of Microsemi.

Episode 17

Part 1

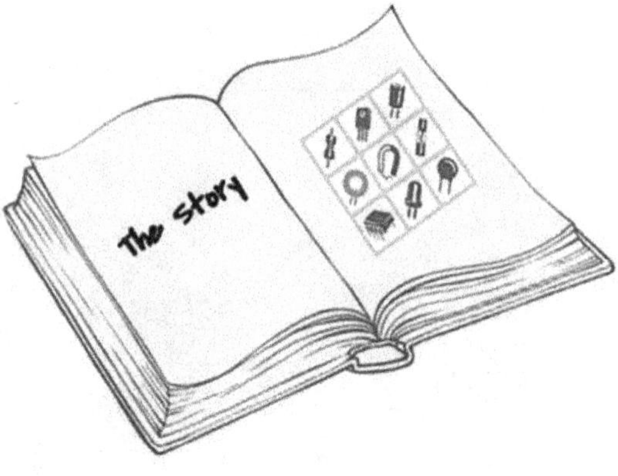
The story

The Story

Foundry Woes

TSMC was founded in 1987 by Morris Chang. At about the same time, I was wrestling with the question of whether or not to join Actel. Morris had been a top executive at Texas Instruments during the period when TI took ownership of the TTL market. (See my episode #8: *Texas Instruments and the TTL Wars*) I have to admit that when I first heard about what TSMC was doing I was unimpressed. I didn't like their prospects at all. (I was wrong.) I was a VP at AMD then. TSMC was anxious to do business with AMD. After having a few discussions with Steve Pletcher, their VP of sales at the time, I formed the opinion that TSMC was ready, willing, and able to do business with pretty much anyone who came to them. (Wrong again!!) Besides TSMC, there was also a lot of foundry capacity available in Japan. Most of the Japanese DRAM manufacturers had overbuilt their fab capacity during the boom of '83 / '84 and were willing to act as foundry sources for American IC houses. So --- my analysis was --- there's plenty of capacity available to fabless semiconductor companies. If I join Actel, foundry will be the least of my worries. (Wrong again. Three strikes and you're out!!!)

There was a problem. There were indeed many different companies with fab capacity. Yes. They were willing to take foundry business. But!!! Essentially none of them were interested in doing business with a custom process. They all had standard processes and wanted to sell those wafers. ----- They didn't want to spend money developing new processes – especially processes that would be used by just one customer. To exacerbate the problem, I hadn't analyzed the process requirements well enough when I was deciding to join Actel. The process we needed was difficult! Extremely difficult!!! Several new steps were going to be problematic, but one of them -- a requirement to build a very, very thin ONO layer (ONO is an oxide - nitride – oxide sandwich-like structure commonly used in DRAMs) with extremely tight control --- was nearly a killer. No one wanted that headache in their fab. To boot, my arguments to the foundries about the huge volumes that we would soon be ordering fell on deaf ears. They just weren't buying it. Despite numerous requests, TSMC always declined our business, politely telling us that they'd be happy to make wafers for us using any standard process they had.

When I arrived at Actel, we had one firmly established foundry relationship -- Data General. DG was a Boston based minicomputer company who had a small, old fab in Sunnyvale.

It was OK for proving out our product, but it wasn't at all a modern, low defect fab. Our yields would be low and our costs high if they were to be our primary foundry. It didn't look like it would be a good long-term relationship for Actel. I wondered what to do about that. While I was wondering what to do about it, they solved the problem for me. They announced that they were shutting down the fab immediately leaving us fabless in the true sense of the word! To be a fabless company with a complex, non- standard process that no foundry wants to run is not a good thing!! ☹ ☹ ☹

We had been having discussions with Matsushita Electric Corporation. (MEC) They liked the idea of programmable products in general and reasoned that it might be good to have access to our technology. They had already made a few test-runs of wafers in their R&D fab for us by the time I got to Actel. Their idea was to get a handle on just how hard the process would be to run. In fact, they had many problems with those test runs, but at least the yields weren't always zero. Of course, we didn't want to give them rights to sell their own FPGAs but were willing to give them rights to use our technology in their ASICs. After almost a (very worrisome) year of negotiating and tinkering around in their R&D fab, they agreed to bring up our process in their production fab if we would let them be our sales rep in Japan. We agreed. All of that sounded good except that the market for ICs was picking up. MEC's fabs were filling up. They had no appetite for spending a lot of effort on our custom process when they had more business than they could handle with their own products and processes. In the end they did bring up our process. A lot of problems had to be solved along the way, but they brought it up. We owe them a lot! But the rules were pretty simple. Don't jerk them around!!! Requests for process tweaks, line holds, splits, etc. would not be tolerated! Those were tough rules that would affect us greatly in the days to come, because our process was far from stable --- it needed a lot of work! And that meant we needed process tweaks, line holds, and splits. How did I get myself into this?!! There seemed to be a new process problem lurking behind every bush --- and there were lots of bushes!!!

Parallel to the MEC negotiations, Actel had also begun negotiations with TI. The general idea was that TI was going to give us some cash and guarantee us foundry capacity. In return they would get second source rights to the first two Actel families. The negotiations progressed well for a while, but then reached an impasse. By the time I came on board, negotiations had ceased. Both sides had walked away from the table. The deal was dead. I joined Actel the morning of December 1, 1988. Early that afternoon Bill Davidow (Who was on our board.) called me and told me that I needed to go to Texas and patch things up in a hurry. He was understating it.

So - the December 1988 status was this: TSMC had turned us down. Data General was shutting down. TI was dead, and MEC was not looking good. We were in trouble!! We had a mask set and nowhere to run it. Cash was beginning to run low. We needed to begin work on another round of venture financing, but I couldn't imagine anyone giving us money if we didn't have a fab committed to building our product. A quick but easy to understand summary?

Looks like we're screwed!!

I wondered if AMD would take me back.

Happily, TI was willing to restart the talks. I flew to Texas with my fingers crossed. Wally Rhines was to be the guy on the other side of the negotiating table. Wally ran the integrated circuit business for TI. I didn't know Wally, but I knew several high-level TI execs by reputation. The short version --- they were very, very, very tough people. Fred Bucy?!! Mark Shepherd?!! Morris Chang?!! Wow. You didn't want to mess with them!! So, my mental picture of Wally was that of a huge man with glowing red eyes, horns, and a tail. When I flew to Dallas for our meeting, I didn't plan on liking Wally. I figured that I was in for a tough day! To my surprise, Wally turned out to be just the opposite. Very bright, but also logical, sensible, and reasonable. After trading a few pleasantries, it might have taken us an hour or so to get things settled. They agreed to be a foundry for us. It took about two years to get the contract drafted and then the process working right, but once we did, they did a good job for us. What a relief. Thanks, Wally!

But, while I'm at it, a little more about Wally Rhines.

In those days, it seemed as though everywhere I went, I ran into Wally talking about digital signal processing (DSP). Technical conferences. Wall Street conferences. Instat. Everywhere! Wherever I went, there was Wally making presentations about DSP. To be honest with you, I had only a vague idea of what DSP was and really no idea at all of what it was good for. To me it looked just like any other microprocessor except that it had an on-board hardware multiplier. "So what?" I thought. The application example he always used was to calculate the five-day running average of the Dow Jones. Again, "So what?"

The fact was that TI had missed the boat in

microprocessors. In fact, everybody but Intel and AMD had missed that boat, but it took some companies a long, long time to figure out that they had missed it. Everybody knew that microprocessors were going to be super important. My sense was that Wally was only trying to assuage his conscience for that miss by talking about some hypothetical but unlikely upcoming DSP surge that TI would own. I was wrong. He was right. He was onto something that it took most of us a long time to figure out. DSP was the missing link when it came to merging the real world (analog) with Moore's Law.

Today DSP is at the heart of all electronic communications -- there's a huge amount of signal processing done in every cell phone and in all the communications links that exist. At the end of the day, DSP was what allowed TI to become a 100-billion-dollar market cap company. Congrats Wally. Hope you were on commission!! Wally later left TI to become CEO of Mentor Graphics.

MEC came up to speed and did a nice job for us for 25 years. TI came up to speed as well and did foundry for us until we bought out their FPGA business in 1995. In both cases, though, the process was always a generation or two behind the state of the art. Xilinx products were always made on state-of-the-art processes.

That was a very serious problem!!!

Episode 17

Part 2

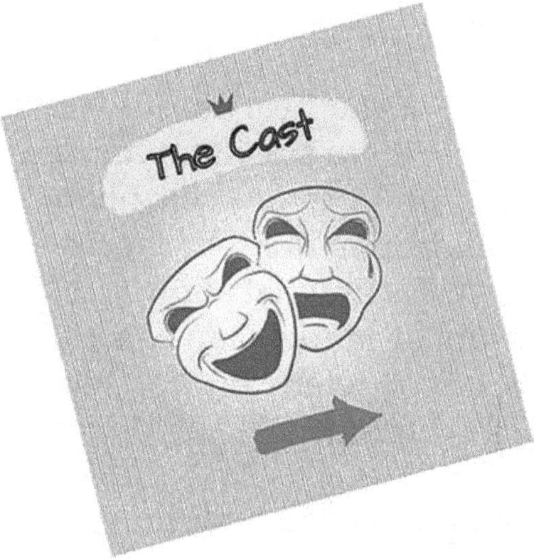
The Cast

The Cast

TSMC

Image 55

We met TSMC earlier in episode #15. The overwhelming majority of IC companies today don't manufacture their own silicon wafers -- instead they buy them from what is called a silicon foundry. TSMC is by far the biggest foundry. In fact, it is now one of Taiwan's biggest companies. Their sales are now running at a $50 billion dollar per year rate and their market capitalization (The value arrived at by multiplying their total number of shares outstanding by the stock market price per share) is about 500 billion dollars (US). For comparison, Intel's market capitalization is about $200 billion as I write this. In episode # 16 you read about Carver Mead predicting the continual "scaling" of the transistor and the benefit that scaling brings. TSMC, along with Samsung, consistently leads the industry in scaling -- they can constantly make the smallest (and hence generally the best) transistors. I really wanted to use TSMC for a foundry. It wasn't to be. They had no interest in developing a custom process for Actel when they were getting all the business they could handle with their standard processes.

Wally Rhines

Image 56

Wally Rhines was born in Pittsburgh, Pennsylvania. He received his PhD in materials science from Stanford University. That was a "chip off the old block" situation. Wally's father had been a professor of materials science at Carnegie Mellon University. Wally spent more than 20 years at Texas Instruments rising rapidly to the position of executive vice president and general manager of TI's integrated circuit business. Wally wanted to get TI into the FPGA business, but they lacked the technology to do it. TI had many wafer fabs and needed FPGA technology. I had FPGA technology and needed a foundry. It seemed like a marriage made in heaven.

Wally moved on to become the CEO of Mentor Graphics - a position he held for 24 years until Mentor was acquired by Siemens in 2017.

Episode 18

Part 1

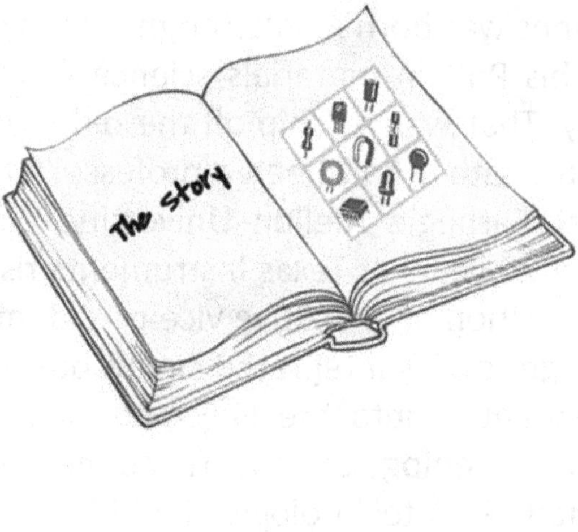
The story

Meeting Andy Grove

The foundry problem continued to plague us at Actel. We had a really complex process! But ---- we needed state of the art feature sizes if we were to compete favorably with Xilinx. TI and Matsushita had been doing a good job for us, but not in fabs with state-of-the-art technology. We were two process generations behind! At two generations behind, we had no chance to compete in density. Xilinx flat out had bigger, more complex FPGAs than we did. Competing in cost and speed was no picnic either. How could we get someone with a state of the art fab to agree to make wafers for us given our non-standard process? One day Bill Davidow and I were brainstorming. Bill said, "Hey, why don't you meet with Intel and offer them a deal. You offer to give them rights to use your programmable technology in exchange for Intel giving you foundry services out of their best fab. That way you'll not only gain a secure foundry partner, but the process you get access to will be the best in the world."

Intel was the technology leader at the time. Bill was right. That deal would have given us access to the world's most advanced technology. A huge win for us!!! But I was skeptical. It wasn't clear to me that Intel would want the rights to our antifuse technology. (Antifuses are explained in episode #16.) Using our custom process with its extra masking, implant, and deposition steps (not to mention the high voltage requirements) would raise the cost of their wafers and hence the cost of their microprocessors. Still - it was worth a try. Bill knew Andy Grove well (by now Andy was the Intel CEO). They had worked together for many years. It was easy for Bill to get me a lunch meeting with Andy.

I called Andy and asked him where he'd like to have lunch. He said, "Right here in the Intel cafeteria" I asked him why he wouldn't rather go to a nice restaurant. His answer: "There's a parking problem at Intel. Not enough parking spots. If I take my car out of the lot to go to lunch, I won't be able to park when I get back'" I asked him if he had a "reserved for the CEO" parking spot. He told me, "no." He told me that the parking protocol was, "The earliest arrivers get the best parking places." So --- if he wanted a good parking spot, the only way to get it was to go to work early and then not move the car at lunch. I liked that system. It's the same one we used at Actel. I told him that I would be happy to meet him at his cafeteria.

Andy Grove was a very, very down to earth guy. When I met him in the Intel headquarters building everyone around him was looking good in their suits, white shirts and silk ties. When Andy came down the stairs though, he was wearing jeans and an old, ugly sweater. We went into the Intel cafeteria. We waited in line with everyone else. He paid for both our meals. Then we found a table that was a little isolated. (There was a circle of mostly empty tables around us. It was a little like the Korean DMZ. Nobody wanted to get too near to Andy.)

Intel had the best fab technology. I wanted to be able to use it! The hard part was figuring out exactly what we could give Intel in return that they would value but wouldn't put them in direct competition with us. I thought I had it figured out. I had prepared a thick binder full of the details of my proposal and all the benefits that would accrue to Intel if they took us up on our offer. I was proud of my work!!! When we sat down, I pulled out the binder and started to open it.

But ... before we got down to business Andy wanted to talk about AMD. AMD and Intel had gone through some very rough legal battles over rights to the Intel processors. AMD maintained that they held certain rights to those processors due to an agreement that the companies signed in 1982. Intel maintained that the agreement was inoperative because AMD hadn't held up their end of the bargain. The legal battle had been very bitter and, in fact, one of the reasons that I left AMD. Andy pretty much hated AMD and everyone who had ever worked there. (But maybe not quite as

much as some of the AMD people hated him ---
it had been a very, very bitter fight). Before I left
AMD, I had been running AMD's microprocessor division --- the group that Andy hated most.
Andy was known to have a quick temper and to
be extremely confrontational. So --- yes! I was nervous! (Here's a good place to insert a joke about
a long-tailed cat in a room full of rocking chairs)

Before we talked about foundry, Andy wanted
to get my views on what the AMD people really thought about what had gone on --- What I
thought about Jerry (Sanders) --- What I thought
about Tony (Tony Holbrook was the president of
AMD during the legal battles) -- What they thought
about the battle. He quizzed me at some length. I
had left AMD several years before I met Andy, so
was mostly able to get away with pleading ignorance. (Of course, I wasn't really ignorant. I knew
exactly what they thought. It wasn't pretty!!!) Finally, he apologized for taking time away from
my meeting purpose and asked me what I had in
mind. I whipped out my massive binder, turned
to page one, and started to take Andy through it.
He reached over and closed the binder.

Andy: "John. In 25 words or less, what is it you
want from Intel?"

Me: "Fab capacity on your advanced line."

Andy: "John. You're a good guy. I like you. So, I'm going to offer you a choice."

Me: "Great. What's that, Andy?"

Andy: "I have a large staff of MBAs who came from really impressive schools. They work on these kinds of proposals for us. They're top notch. If you'd like me to, I'll give them your proposal and ask them to study it thoroughly and provide a well-reasoned, written response. That will probably take them a month or so. I'm quite sure their answer will be no. The other option is that I can tell you no right now and save you from having to wait a month. Which way would you like to go on this?"

There was no beating around the bush when you were dealing with Andy Grove!

Episode 18
Part 2

The Cast

The Cast

Andy Grove

> Success breeds complacency.
> Complacency breeds failure. Only
> the paranoid survive.
>
> — *Andy Grove* —
>
> AZ QUOTES

Image 57

You've already met Andy Grove twice. Once in the episode about the early days of Intel (#5) and once in the episode about the IBM PC (#12). Andy was a hard driving, in your face manager of people. He could be (And indeed often was) extremely intimidating. He wanted the people who worked for him to manage the same way. He wrote two books on management: "High Output Management," and "Only the Paranoid Survive." Both were really big sellers in the high-tech world. Intel (under Andy) and AMD (under Jerry Sanders) spent a decade or so engaged in an extremely antagonistic legal battle over the rights to the Intel processors. Both sides hated the other side. I'm sure Andy had plenty of things he'd rather have done than to meet with me -- a known AMD antagonist. But -- he did meet with me. I appreciated that. I ran into Andy fairly frequently after our first meeting - often at parties, conferences, or other industry affairs. He was always cordial. The AMD side of me hates to admit it, but I grew fond of Andy Grove.

Bill Davidow

Image 58

You've already met Bill Davidow twice. Once in the episode about my move to Actel (#16), and once in the episode about the IBM PC (#12). Bill was a highly trained engineer, but his true love was marketing. His first book, Marketing High Technology, put a new spin on traditional marketing techniques and arrived at a new set of techniques as well as a new way to think about the marketing of high-tech products. It was required reading for marketing executives at Actel and many other companies. Bill had worked with Andy Grove for many years. They respected each other greatly.

Episode 19

Part 1

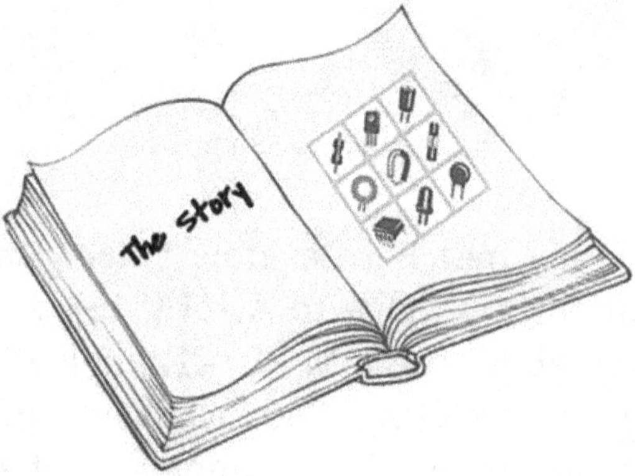
The story

The Story

Going Public

In 1990 Xilinx notified us that they believed Actel was infringing a patent that had just been issued to them. My immediate thought – "The patent system is all screwed up!" Actel had been developing our product for five years. We had been shipping it for a year and a half. During all that time, we were totally unaware that there was or ever would be such a patent -- there was no way to know that Xilinx had filed for a patent or what was in the filing. According to patent law, patent proceedings are secret until the patent is finally issued. Sounds like a great law, right? But when the Xilinx patent did issue ---- BAM!! We'd been blindsided!! The "normal" way to handle this situation is to take a license -- to agree to pay a royalty to the owner of the patent. In this case there was a problem with that. Bernie Vonderschmitt (The CEO of Xilinx) didn't want to give any licenses. I can't really say that I blame him for wanting to keep the FPGA market to himself. It was hard to get mad at Bernie. He was a classy guy. Yet, if it had gone to court and Xilinx had won, they would have been completely within their legal rights to enjoin Actel from shipping our FPGAs. That would have forced us to close our doors – to shut Actel down.

I spent the better part of three years trying to solve this potentially very, very serious problem.

In those days the rule of thumb was that a start-up could go public once they had achieved two consecutive profitable quarters. We reached profitability in 1990 and were hot to go public. We selected bankers -- Goldman Sachs was our "bulge- bracket" bank -- and started preparation. Then, Goldman's lawyers started to get cold feet. What if the patent battle couldn't be settled? If the worst case came to pass --- that is -- we went public, took money from new shareholders, and then had to shut our doors because we had lost the patent suit --- the lawsuits would be everywhere. Eventually we had to put the IPO on hold. (IPO = Initial Public Offering- also known as "going public".)

I'm not sure that we could have settled the Xilinx patent suit. Bernie hadn't shown any interest in doing so. But then a good break came our way. Khalid El Ayat found that the early Xilinx parts didn't use what we called segmented routing but that their later family (the 4000 family) did. One of our original patents had some claims dealing with segmented routing techniques (thanks Abbas El Gamal, Jon Green and team). That allowed us to file a counter suit against Xilinx. Calmer minds prevailed. We settled the suits peacefully.

Note: Xilinx later sued Altera for infringing that same patent. They weren't able to settle and eventually went to trial. My analysis was that Altera did infringe, but that the patent wasn't valid over prior art. The jury eventually ruled that the patent was indeed valid, but that Altera didn't infringe. In my mind, wrong on both counts. That's the reason that you never want to go to a jury trial with a technical issue. The jury won't understand it and has an excellent chance of getting it wrong. This case was very technical. I doubt that the jury members understood 10% of what they were being told.

Once the Xilinx suit was settled, our IPO efforts cranked up in full force. Everything was looking good. Then we hit another snag.

We had been selling 1.2-micron material, but we needed 1.0-micron product if we were to be cost competitive. The process was ready to go, or so we thought, but the first production runs to come out had terrible yields. If we couldn't run the 1.0 process, we couldn't compete well cost-wise. That would have to be disclosed in the IPO documents. We had to put the IPO on hold for a second time. Then, we cracked the code. We figured out the problem and fixed it. Thanks Steve Chang! Thanks, Esmat. (Esmat Hamdy was Actel's vice president of technology development at the time) The IPO was back on.

IPOs are hard work. Huge amounts of effort go into putting together the S1. (The SEC's form S-1 is the master document that is used to communicate with them.) Everything that could end up mattering to a prospective shareholder must be disclosed and explained in the S1. The explanations had to be right. If the stock went down for any reason other than the market's fickle nature, you could expect a lawsuit. The point of the suit would be that you hadn't properly disclosed the risk of some aspect of the company's business and that if you had, the shareholders wouldn't have bought the stock and hence wouldn't have lost their money. I spent many, many, many hours writing those documents. It took a long time and a lot of effort by a lot of people to get the documents ready! Those documents are a big, big deal! Luckily for me our CFO, Rich Mora, had been through it before. He knew what had to be done. He understood how to do it. I would have been in deep trouble without Rich.

Once you have the documents done, they must be approved by the Security Exchange Commission. The SEC folks are tough. They invariably find fault and demand improvements. More time. More effort. Once the SEC has given their approval, you can start the "road show." The point of the road show is to meet with the people who are about to invest and teach them the basics of the company's business. The road show is grueling. You present the company pitch over and over and over again.

Breakfast presentations. Morning one-on-ones. Lunch presentations. Afternoon one-on-ones. Dinner presentations. Then, you head to the airport, fly to a new city, and do it again. One day we had a breakfast meeting in Minneapolis, a lunch meeting in Chicago, a dinner meeting in Madison, and then flew to the east coast where a full day's work was set up for the following day. When we landed, Dave Courtney (Our Goldman Sachs guy) had a message that our afternoon meeting the next day had been canceled. That would mean that we had a couple of hours on our hands. I suggested that we go see the Liberty Bell -- I had never seen it. Dave thought that I was really stupid! "John, the Liberty Bell is in Philadelphia. We're in Boston." Actually, I knew where the Liberty Bell was. I just didn't know where I was. All the cities had become a blur by then.

A few days later in New York City we got on a plane at Kennedy Airport and prepared to head home. We were done!!! It was a Friday afternoon. We would celebrate over the weekend. Monday morning when NASDAQ opened there would be a new listing: ACTL. Has a nice ring to it, don't you think? ACTL!!! Just as I was going through security check, someone from Goldman brought me a fax that they had received a few minutes earlier. I boarded the plane, took a sip from the glass of wine that I'd already treated myself to and opened the envelope. It was a letter from a struggling start-up accusing us of infringing a patent of theirs. I almost spit out the wine!!! When I took a quick look at the patent, it wasn't clear that it really applied to us. But -- their timing was impeccable -- clearly calculated to get some quick cash from us. The patent would constitute a new "risk factor" that should be disclosed to all potential investors. To do that, though, would mean putting the IPO on hold for a third time. That would mean redrafting and reprinting all of the documents, waiting a month or so for the new info to disseminate, going through SEC approval again and then doing the road show all over again. Devilishly clever timing, don't you think? But also, in my mind, completely without class. We were supposed to "price" and open the market Monday morning. The weekend when I planned to relax and celebrate was shot. We sorted it out that weekend. I don't remember any details, but I'm sure it involved a tidy payment to the very clever offended party. What I do remember is being really relieved that we had sorted it out.

Monday morning, August 3 1993, when they rang the opening bell*, Actel stock was trading publicly right along with the likes of Apple and Intel!! We opened at $9.50 per share. During the day we ran up to $13.00. We were a public company.

Actually, they don't ring a bell at NASDAQ. They just push a button. It's the New York Stock Exchange that rings a bell. I found that out in 2003 when I was invited to open the market as we celebrated our tenth anniversary of being a public company. I gave a little speech, waited a couple of minutes, and when they gave the sign, reached out and pushed a small black button.

So easy that even a CEO could do it!

Episode 19

Part 2

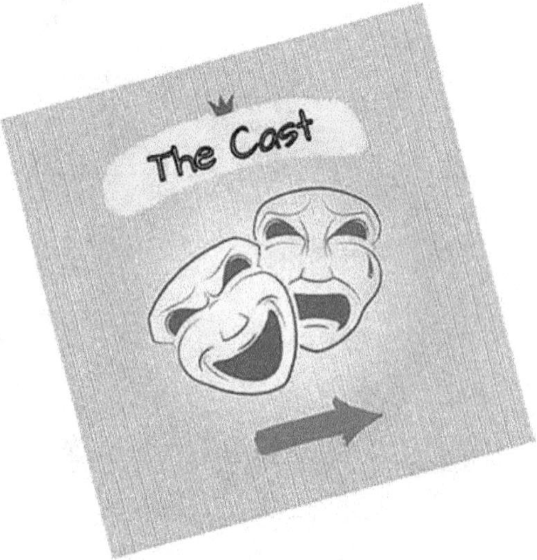
The Cast

The Cast

NASDAQ and Actel

Image 59

NASDAQ is the stock exchange of choice for almost all high tech initial public offerings (IPOs). NASDAQ is de-centralized -- the traders live all over the world and are connected with each other by a computer network. There is no massive central trading floor where all the traders run around frenetically buying and selling stocks. In simple terms, NASDAQ is an electronic stock market. The headquarters are located in the middle of Times Square in New York City. When a company is newly listed on NASDAQ and then again on the tenth anniversary of that listing the company name and stock symbol are displayed on a huge electronic sign on the NASDAQ headquarters building. After ten years of being listed the company is invited to visit New York and to "open" the market trading that day. This is a shot of what could be seen by anyone walking up and down Broadway in New York City on August 1, 2003 -- the tenth anniversary of Actel being listed on NASDAQ.

The S-1 Document

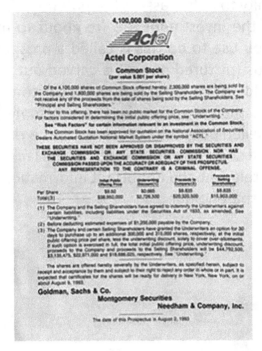

Image 60

Form S-1 is the document that must be provided to the Securities Exchange Commission (SEC) before going public. The SEC makes an assumption that private companies are bought and sold by sophisticated investors who have a thorough understanding of the transactions that they are making. (Often true. Not always.) On the other hand, the SEC presumes that many of the people who buy and sell stocks on the open market have no such sophistication. (Definitely true) The idea behind the form S-1 is to spell out the business operations and financial status of the company as well as the associated risks in simple, easy-to-understand terms. The goal is that non- sophisticated investors have the opportunity to better assess the potential risks and rewards of the investment they're about to make. The vehicle you use to do this is called a prospectus. The prospectus (pictured above) is carved out of the S-1. This should not be taken lightly! If your company goes public and then the stock falls due to something that wasn't spelled out in the prospectus as a risk, you can count on a shareholder lawsuit which claims that you withheld relevant and important information from the potential shareholder.

Bernie Vonderschmitt

Image 61

You met Bernie Vonderschmitt in the episode about the time when I was joining Actel. (#16) Bernie was a real gentleman!! But ... as CEO of Xilinx, he had a job to do. Part and parcel of that job was defeating their competition. No better way to do that than to use a patent to shut them down. Xilinx had filed for some very broad patents covering the concept of an FPGA (although they didn't call it an FPGA in their filings). By law a patent application is kept secret until the patent is issued. Their first relevant patent was issued four years after Actel was founded. During that four years, Actel had no way to know that there would be a relevant patent or what would be in it. Bernie wasted no time in asserting the patent against Actel once it had been issued.

Rich Mora

Image 62

Rich Mora received a bachelor's degree from University of Santa Clara. He spent the early part of his career as an auditor for Price Waterhouse Coopers – a major auditing firm. Then he decided that rather than audit a company and tell them what they might be doing wrong, he'd rather work at the company and show them how to do things right. Instead of opting to work at a large, public company, he chose small, private companies. The thing about small, private companies is that they don't always stay small and private. Sometimes they grow to be big. Sometimes they go public. In Rich's case, he's been involved in five different IPOs and S-1 registrations. I think he may have the Guinness World Record for registrations.

So – yes. Rich knows a thing or two about form S-1.

Episode 20

Part 1

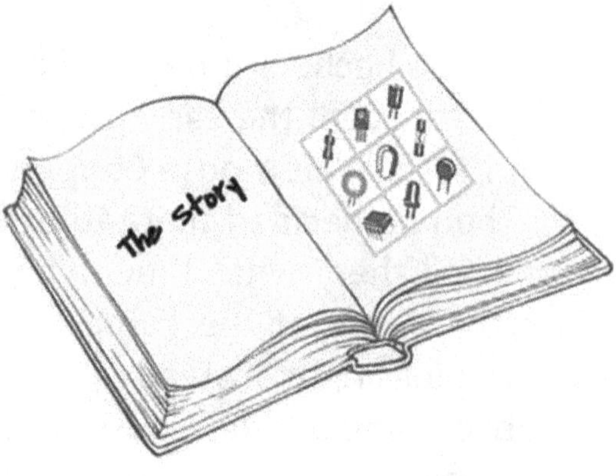
The story

The Story

Crashing the Mars Rovers!!!

In early 2003 Actel announced a new product family: RTSX-A. It was a family of antifuse FP-GAs aimed at the satellite market. Customers had known for a long time that it was coming and there had been prototypes available for many months. "Space customers" loved the product. Our FPGA was non-volatile. Our competitors' FPGAs were volatile. The designers of satellite and other "outer space" systems prefer non-volatile parts. (The terms volatile and non-volatile are explained in episode #6) This product was going to be a big win for us! One of the first programs to use the product was the Mars Rover program. That was a huge, huge design win for us!

The Mars Rover program was the most exciting thing in the press while it was happening, but that was almost twenty years ago. For those of you who don't recall, here's a quick summary. There were four Mars Rovers made: two went to Mars and two lived in a huge sandbox in Pasadena which allowed scientists to emulate conditions on Mars. The two that went to Mars were named the Spirit and the Opportunity. NASA's plans were that the Rovers would live for three months before succumbing to the treacherous Mars environment. (On Mars you'll find staggeringly low temperatures, an unbreathable atmosphere and horrific dust storms that last for months mixed in with a bit of deadly radiation. It's not a nice place.) The Spirit launch was on June 10, 2003. It would take the Spirit about eight months to make the trip to Mars. The Opportunity was launched a month later.

Shortly after the launch we got reports from potential customers of a few RTSX-A burn-in failures in their labs. What's burn-in? It's a test that assures the user that a part that starts out good will stay that way after being in actual use for some time. Testing parts for one week at extremely high temperatures is the normal way to assure that they'll last a long time at normal temperatures. We hadn't seen failures with our first tests, but when we got those reports from our customers, we looked harder. When we looked harder, we saw some failures on occasion. We would burn-in around 100 units at a time. Sometimes we would get 1 or 2 failures. Sometimes none.

A small number of failures is worse than it sounds! A typical satellite would cost in the neighborhood of 100 million dollars in those days. The Mars Rover project vehicles cost much, much more than that. Satellites can't be repaired. If one

IC fails, the cost of the entire project may well be flushed down the toilet. Worse, there wasn't just one RTSX-A part in each Rover. There were, as I recall, 38. That meant if there was a 1% chance that one particular Actel part would fail, there was a 38% chance that one of our parts somewhere in the Rover would fail. (My math isn't quite correct here, but you get the point). Both Rovers had already been launched. Oh-oh!

It was clear something wasn't quite right, but we couldn't figure out what. Worse, the Mars Rover program was by no means the only program planning to use our parts. There were many others! And many of those were US government satellites critical to the defense of our nation. Word got around the space community. Many customers weren't sure if they dared to launch their satellites. They looked to us to tell them it was OK. We couldn't. We just didn't know. We couldn't figure it out. A few skeptics thought we were covering something up. (A conspiracy theory? Don't you love conspiracy theories?) We weren't. It was a very tricky problem. We were working hard on it, but we just didn't understand what was going on.

Then, I got a call from Dr. Bill Ballhaus, the CEO of Aerospace Corporation. Aerospace Corporation operates a federally funded research and development center. They provide technical guidance and advice on all aspects of space missions to military, civil, and commercial customers. Dr. Ballhaus asked me to come to the Aerospace headquarters in El Segundo to discuss "the reliability problem with Actel FPGAs."

I, of course, accepted the invitation, but if you had offered me a choice of going to this meeting or getting a root canal on every tooth, my oral surgeon would be a richer man today. The invitation appeared to be for a one-on-one meeting between me and Dr. Ballhaus. I planned on going by myself, but our Vice President of Technology, Esmat Hamdy, saw it differently. He didn't quite trust my technical savvy. He thought that, if the meeting turned out to have a lot of technical content, I might not be able to answer all the questions well. So --- Esmat insisted on coming with me. Bless his heart!

We flew to LAX and then took a quick cab ride to Aerospace. It's about a mile from the airport. A secretary led us to a conference room --- except it was more like a sports arena. There was a long rectangular table that probably sat 15-20 people. Then there was an aisle circling those seats. But on the other side of the aisle, there was an elevated set of chairs circling the table below. There were maybe another 20-25 chairs in that set. In total I would guess 30 or 40 chairs. All but four were full. Not full of just anybody --- but full of PhDs. Full of experts in any aspect of integrated circuits that you could think of. Full of technical wizards who all had at least double my IQ. One of the empty chairs was for Dr. Ballhaus. One for Esmat. One for me. We took our chairs and were ready to start the meeting, but Dr. Ballhaus said that we'd have to wait. The last chair was for a high-ranking Air Force general who had invited himself to the meeting. The general was running late. When he got there, there was plenty of bowing and scraping done.

They hammered into me just how important this was. The Rover program cost about one billion dollars. The Rovers had been launched. There would be no calling them back. No fixing them. If they went bad, that would be a billion dollars flushed. But the Mars Rover project was by no means the only satellite project with plans to use Actel. As you've already heard, there were several military satellite programs related to our national defense as well. Those folks were even more worried than the Mars Rover people. They were not happy campers!!! As you would expect --- the general had zero interest in putting up military satellites critical to the nation's defense that were likely to fail!!!!! (That unhappiness earned me an invitation to meet later with Peter Teets, the Undersecretary of the Air Force, in the most secure area of the Pentagon. Mr. Teets wasn't a happy camper either, but that's another story.)

Back in Mountain View, we were breaking our picks. We didn't always see failures. When we did see them, there weren't many. That makes the problem harder to solve. We suspected what we called the programming algorithm. The oversimplified explanation of programming an antifuse is this -- put a high voltage across it, the dielectric will rupture and the antifuse will be a conductor for the rest of time. Sounds simple! In fact, it's much trickier than that. There are a lot of knobs to turn. How high should the voltage be? How much current should flow through the fuse? How many times should you repeat what you've done? How long should you apply the voltage? How much should you "soak" the fuse? We would twist some of these knobs, come up with a new algorithm, and voila. No failures!!! Problem solved, right? Wrong!!! The next time we'd run exactly the same test using the same new algorithm but there would be one or two failures. We were completely perplexed!

The first Rover (the Spirit) was launched prior to suspicions that we might have a reliability problem. Then came the reliability worries. And then, around New Year's when the reliability concerns had become rampant, the Spirit reached Mars. It was a big deal in the press.

It was on the front page of every newspaper. Boy. Did we ever feel good! The Spirit was working perfectly!!! .... But then --- one week later ----

The Spirit went bad. That was on every front page too, but in bigger letters. Was it the Actel parts? We didn't know, but in my judgement ... it could well have been. I was terrified! I could picture the headlines when it was determined that Spirit failed because of an Actel part. I could picture the lawyers lining up to file lawsuits against us. I could picture the process servers skulking in the bushes waiting to spring out and serve me with subpoenas. It wasn't pretty!! In fact, it was really, really ugly!! When I went home after work that night, I opened a bottle of wine and drank the whole thing by myself. I like wine ...... but not a whole bottle. My advice? -- Don't do that!!! It didn't work out well!!

Luckily, I was wrong. The Actel parts were not at fault. After a week or so, NASA figured it out. It was a software problem that was fixable by up-loading new software. They fixed the Rover, and independently we tracked down our problem and fixed it. To my knowledge Actel (Now part of Microchip Corporation) has never experienced a failure in space.

Scientists had planned for the Rovers to live for three months. When did they actually die? The Spirit lived for 6 years. The Opportunity 14.

For the sake of the curious-minded, here's what we were up against. To properly test an FPGA you would typically program 10,000 or so antifuses in it. So – when you tested 100 FPGA units you were actually testing 10,000 times 100 = one million antifuses. When one FPGA out of one hundred failed, that meant that one antifuse out of a million was bad – somehow different than the other 999,999. The brute force approach would be to inspect all million before programming and note any differences no matter how small. After you saw which antifuse didn't work right, you'd know what you were trying to fix. The problem with that? The antifuses were buried deep inside the silicon. You couldn't see the antifuse without tearing the FPGA apart. Then you couldn't program it. How about doing the inspection after programming and life test? Then you'd only have to look at the bad antifuse, right? Nope!! To the first approximation, programming an antifuse is akin to blowing it up. By programming it, you're destroying the evidence.

This problem took 20 years off of my life. (And 3 0 years off of Esmat's)

Episode 20

Part 2

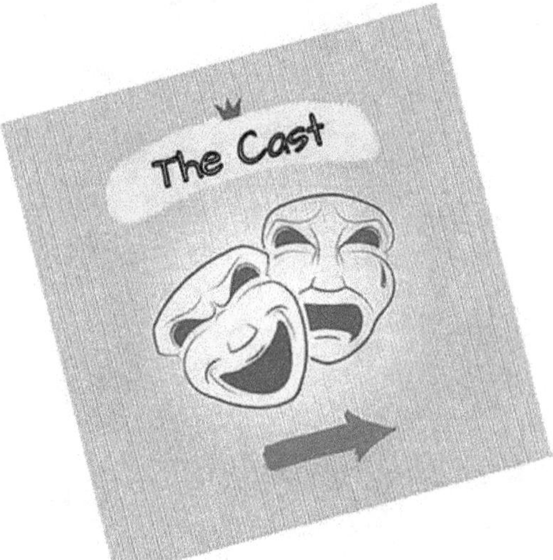
The Cast

The Mars Rovers

Image 63

Before the Mars Rovers were launched, I gave a talk to some of the scientists at NASA. It seemed like the thing at the top of their minds was that "Mars is tough." Many of the missions aimed at learning more about Mars over the years had been plagued with technical difficulties. It was as though someone had fashioned a Mars voodoo doll and continually poked it with needles whenever a Mars mission grew near. NASA was nervous about the upcoming Spirit and Opportunity launchings. I, on the other hand, was elated. Actel was about to make a name for itself. We would soon be known as the IC company most responsible for the tremendous technical victory that the United States was about to achieve!!!!

Or would we?

Peter Teets

Image 64

Peter Teets was born in Colorado. He received a Master of Science degree at MIT. He started work as a design engineer at Martin Marietta and eventually rose to the position of CEO of Lockheed Martin (Lockheed and Martin Marietta had merged in 1995). At his request I met with him and a few of his generals in early 2004. The generals said nothing. I think they were afraid to speak. When I met Teets, he was serving in two capacities: Undersecretary of the Air Force and head of the National Reconnaissance Office. He simultaneously reported to three different bosses: The Secretary of the Air Force, the Secretary of Defense, and the head of the National Intelligence Agency. He's a very intelligent man. He's also a very serious man!! To put it mildly, Teets didn't like the idea that his critical defense satellites might all fail because of a problem with the Actel chip.

I had to change underwear immediately after the meeting.

Esmat Hamdy

Image 65

Esmat Hamdy was born in Egypt but moved to Canada when he was still in college. He received a PhD in electronics from the University of Waterloo in Waterloo, Canada. He was one of six founders of Actel. 35 years later, he still works at "Actel" although Actel is now a part of Microchip Technology. Esmat started out at Actel as the antifuse expert, but rapidly ascended to the position of vice president of operations and technology development.

Antifuses are hard!!!! Antifuses are nasty!!! There are more things that can go wrong with an antifuse than I ever dreamed of until I got to Actel. Then, all of them did!! Esmat and his team (which included John McCollum -another founder of Actel) tackled every one of those problems and one by one, solved them all. But no one was having any fun along the way. I love Esmat Hamdy.

But Esmat, I hate antifuses!!!

Epilogue

Somewhere around the year 2000 our board of directors asked the question, "John, how long do you plan to stay on as CEO?" My answer: "I'll retire no sooner than my 65th birthday and no later than my 66th." When you're 55, 65 seems really old, doesn't it? Well, ten years later along came my 65th birthday. January 20, 2010. Funny thing. By then 65 didn't seem so old. On the other hand, I had learned a lot in that ten-year window. I learned some things that have proven very helpful and others that I really never wanted to know. For example: Flash FPGAs are hard too!! And .... Not all activist shareholders are nice people!! And ... The worst day on a Mediterranean cruise is better than the best day in court. All in all, it was the right thing to do. We released an 8K (that's the document that public companies use to disclose relevant information.) saying that we were beginning a search for a new CEO and that we planned to complete the search and appoint a new CEO within a year.

We spent several months on the search. There were some ups and downs – but bottom line, we made zero progress. We were back to square one. Then came a surprise. We received an unsolicited and unexpected offer to buy us out from Microsemi -- an Orange County semiconductor company that I was barely aware of. We discussed it at length.

Why would we be interested in selling? We were in the middle of trying to transform ourselves from an antifuse company to a flash company. I firmly believed that it could be done (In fact, it was done!!! See episode # 16, "From AMD to Actel"), but it was obvious that it was going to take a long time before we finally started to see the results on our bottom-line. The bottom-line matters!! Shareholders want high stock prices!! And --- for a company our age, the stock price is determined by the profits you're making ---- that is ---- by the bottom-line.

Image 66

Dan McCranie (Who we had recently appointed chairman of the board) went to Orange County and met with the Microsemi CEO, James Peterson. Dan is good at everything he does. He started out as an engineer. He moved to sales. He served on nine different boards of directors.

He was always good. As a negotiator he's good as well. After some negotiations, Peterson offered a share price that was higher than we believed we would be able to command consistently in the foreseeable future. After a few vigorous board discussions, we decided that we owed it to our shareholders to accept the offer. We did.

On November 3, 2010 Actel became part of Microsemi Corporation and I rode off into the high-tech sunset .... unemployed for the first time in 42 years.

A parting thought. It was known that the Soviet Union was working on semiconductor technology after World War II. Japan was as well. What if they had won the race? The development of the integrated circuit not only gave Silicon Valley its place on the socio-economic map of the United States, it also gave the United States its place on the map of today's world.

Recommended Reading

My junior year at Berkeley I took EE105. It was a required course for all electrical engineers. That was the course that taught transistor physics (Mostly bipolar junction transistors – not MOS). The professor was top notch -- one of the best in the world. Still, I walked out of most of the lectures not really understanding what had been said. When the course was over, I still didn't feel like I understood how a transistor worked. My attitude was more or less, "If I never see another transistor it'll be too soon!!" (Later in life I felt the same way about antifuses). Well -- I ended up seeing a lot of them. What I have a harder time with today is understanding; "How did we go from being barely able to make a single one of those little 'transistor things' to being able to crank out millions of circuits each of which contains ten billion of them?!!!" I've been a voracious reader on the subject. There are any number of great books. Here are my favorites.

The Microchip Revolution. *By Luc Bauer and Marshall Wilder. Gets into the technical details of semiconductor history. Written by two men who spent 45 years each developing that technology. PS. Thanks to Marshall Wilder and his wife Gale for proof reading this book!*

Spinoff. *By Charles Sporck. The early years of Fairchild and National as seen through Charlie Sporck's eyes. A fun read!*

The Code. By Margaret O'Mara. Looks at 80 years of technological progress including, but not limited to, the IC revolution. Goes on to cover computers, Microsoft, the internet, etc.

Fabless. By Daniel Nenni. Tells the story of the fabless model taking over the industry.

Makers of the Microchip. By Christophe Lecuyer and David Brock. Uses Jay Last's laboratory notebooks to get the inside story of the early days of Fairchild. Full of fascinating detail. A must read for true lovers of IC history!!

Microcosm. By George Gilder. Starts with the development of semiconductors but goes on to look at some economic and philosophical ramifications.

The Big Score. By Michael Malone. This book has some fun stories about Fairchild, National, Intel, and AMD. Goes on to cover much of the computer revolution.

Silicon Destiny. By Rob Walker. Rob tells the story of how the application specific integrated circuit came to be. He should know. He's the one who did it. (Actually, he was one member of a fabulous team.)

History of Semiconductor Engineering. By Bo Lojek. A top-notch technologist takes us from Bell Labs to present day with a bent that's partly technical and partly philosophical.

The Man behind the Microchip. *By Leslie Berlin. A biography of Robert Noyce*

Broken Genius. *By Joel Shurkin. A biography of William Shockley.*

The Chip. *By T.R. Reed. Covers the entire history of semiconductors and has lots of detail regarding the Kilby-Noyce "competition".*

Crystal Fire. *By Michael Riordan and Lillian Hoddeson. By far the most detail on the development of the transistor that I have ever seen. Barely gets into the integrated circuit.*

About the Author

John East retired from Actel Corporation in November 2010 in conjunction with the transaction in which Actel was purchased by Microsemi Corporation. He had served as the CEO of Actel for 22 years at the time of his retirement. Previously, he was a senior vice president of AMD, where he was responsible for the Logic Products Group. Prior to that, Mr. East held various engineering, marketing, and management positions at Raytheon Semiconductor and Fairchild Semiconductor. In the past he has served on many boards of directors and advisory boards for companies involved in various aspects of the high-tech market. He currently serves on the boards of directors of SPARK Microsystems – a Canadian start-up who designs and markets high speed, low latency, low power radio chips — and Tortuga Logic — a Silicon Valley start-up with expertise in detecting and correcting hardware security issues. He is presently an advisor to Silicon Catalyst — a Silicon Valley based incubator actively engaged in fostering semiconductor-based start-ups. His education and work histories are chronicled by the Computer History Museum at:

https://www.youtube.com/watch?v=naqKYxxqLss

as well as in the Stanford University Library / Silicon Genesis at:

https://exhibits.stanford.edu/silicongenesis/catalog/yp332mx7612

Mr. East holds a BS degree in Electrical Engineering and an MBA both from the University of California, Berkeley. He has lived in Saratoga, California with his wife Pam for 49 years.

Gallery of Images

Episode 1 Part 2

Image 1- Shockley Semiconductor

Photo: The Arnold and Mabel Beckman Foundation
License CC BY-SA 4.0

Image 2- Origin of Silicon Valley

Image from the John East Collection

Image 3- Fairchild Semiconductor

Image from: https://www.wikiwand.com/en/Fairchild_
Semiconductor License CC BY-SA 4.0

Image 4 -and 4a- John and Pam East

Image from the Pam East Collection

Image 5- Les Hogan

-Image from: https://engineering.lehigh.edu/alumni/
clarence-hogan CC BY-SA 4.0

Episode 2 Part 2

Image 6- Charlie Spork

interviewed by Floyd Kvamme, on 2014-11-21
X7310.2015 YouTube License CC BY-SA 4.0

Image 7- John Carey

image provided by the computer museum in San Jose,
CA

Image 8- Traitorous 8

Image by Chipsect.com licensed under CC BY-SA 4.0

Episode 3 Part 2

Image 9- William Shockley

From a 1974 interview with William Buckley Jr. on Firing Line-YouTube CC BY-SA 4.0

Image 10- John Bardeen, William Shockley, and Walter Brattain

Image by At&T CC BY-SA 4.0

Image 11- The Nobel Prize Celebration

Image from CoreCompass.com licensed under CC BY-SA 4.0

Episode 4 Part 2

Image 12- Robert Noyce

Image from: Azquotes.com CC BY-SA 4.0

Image 13- Jack Kilby

Image at https://www.thefamouspeople.com/profiles/jack-kilby-6393.php license CC BY-SA 4.0

Image 14- Jack Kilby and Robert Noyce

Image from IDG.com licensed by CC BY-SA 4.0

Episode 5 Part 2

Image 15- Gordon Moore

Image from: Azquotes.com CC BY-SA 4.0-

Image 16- Andy Grove, Bob Noyce, and Gordon Moore

PHOTO CREDIT: COURTESY OF INTEL CORPORATION CC BY-SA 4.0

Episode 6 Part 2

Image 17- The Microprocessor

Image by Sandy Schaeffer for the National Science Foundation licensed under CC BY-SA 4.0

Episode 6 Part 2 (Cont)

Image 18- Federico Faggin

Photo by Elvia Faggin. licensed under CC BY-SA 4.0

Image 19- Eli Harari

Photo: computer history YouTube licensed under CC BY-SA 4.0

Episode 7 Part 2

Image 20- Fairchild Assembly

Image Dated 1964 Curtesy Mercury News Archives CC BY-SA 4.0

Image 21- Walker's Wagon Wheel

Carolyn Caddes papers and photographs, circa 1980-2015, licensed under CC BY-SA 4.0

Episode 8 Part 2

Image 22- TTL Gate

Image from John East Collection

Image 23- Tom Longo

Photo: Public Domain licensed under CC BY-SA 4.0

Episode 9 Part 2

Image 24- Wilf Corrigan

Image from: almostdailybrett.wordpress.com licensed under CC BY-SA 4.0

Episode 10 Part 2

Image 25- Charlie Sporck

Image from LinkedIn by Klaus Leutbecher licensed under CC BY-SA 4.0

Image 26- Pierre Lamond

Image from eclipse at crunchbase.com licensed under CC BY-SA 4.0

Episode 10 Part 2 (Cont)

Image 27- Traitorous 8

Image from WAYNE MILLER/MAGNUM PHOTOS

Episode 11 Part 2

Image 28- Jerry Sanders

CC by SA 2.0 AndyTru flicker

Image 29- T J Rogers

Image from Alchetron.com licensed under CC BY-SA 4.0

Image 30- Jayshree Ullal

Photo; by permission of siliconANGLE CC BY-SA 4.0

Image 31- Jen-Hsun "Jensen" Huang

NVIDIA Corporation Jen-Hsun Huang Head-shot CC BY-SA 4.0

Image 32- Lisa Su

Image by AMD Global CC BY-SA 4.0

Episode 12 Part 2

Image 33- Paul Indaco

Image from Semiwiki.com cc BY-SA 4.0

Image 34- Paul Otellini

Photo by Tech Central CC BY-SA 4.0

Image 35- Andy Grove

Photo by ChiefExecutive.net CC BY-SA 4.0

Image 36-Bill Davidow

Image from Azquotes.com CC BY-SA 4.0

Image 37- David House

Image from Crunchbase.com CC BY-SA 4.0

Episode 13 Part 2

Image 38- Steve Jobs

Image by: :AZQuotes.com CC BY-SA 4.0

Image 39- Steve Jobs and John Sculley

Photo by: Stevejobsblogspot.com CC BY-SA 4.0

Image 40- Apple Computer

Photo by: Werd CC BY-SA 4.0

Image 41- Steve Wozniak

Image by Gage Skidmore CC BY-SA 2.0

Episode 14 Part 2

Image 42- NeXT Computer

Image at hertimehascome.com W, Taylor CY Nueva School CC BY-SA 2.0

Image 43- Steve Jobs and NeXT

Image from Vintage Photos on twitter @NotableHistory Licensed by: CC BY-SA 4.0

Image 44- Rich Page

Image from The John East Archives

Image 45- Gil Amelio

Image from azquotes.com licensed under CC BY-SA 4.0

Episode 15 Part 2

Image 46- TSMC/Morris Chang

Photo by Shinya Sawai CC BY-SA 4.0

Image 47- Wafer Fab

Photo by: Wikichip.org CC BY-SA 2.0

Episode 16 Part 2

Image 48- Gate Arrays, FPGAs, and ASICs

Image provided by Microchip

Image 49- John Birkner and HT Chua

Image from Monolithic Memories, Inc. CC BY-SA 4.0

Image 50- Ed Zschau

Photo by: The trusties of Princeton University CC BY-SA 4.0

Image 51- Bill Davidow

Image from YouTube "Bill Davidow: Overconnected" CC BY-SA 4.0

Image 52- Carver Mead

Photo by: :AZQuotes.com CC BY-SA 4.0

Image 53- Bernie Vonderschmitt

Image courtesy of Xilinx, Inc

Image 54- Steve Sanghi

Image from Facebook.com CC BY-SA 4.0

Episode 17 Part 2

Image 55- TSMC

Image from: mc.tsmc.com CC BY SA-4.0

Image 56- Wally Rhines

Image from European Business Press CC BY-SA 2.0

Episode 18 Part 2

Image 57- Andy Groves, Revisited

Image from: https://www.azquotes.com/quote/118559 licensed under CC BY-SA 4.0

Episode 18 Part 2 (Cont)

Image 58- Bill Davidow

Image from: https://economictimes.indiatimes.com
Licensed by CC BY-SA 4.0

Episode 19 Part 2

Image 59- NASDAQ and Actel

Image from the John East Archives

Image 60- The S-1 Document

Image from the John East Archives

Image 61- Bernie Vonderschmitt

Image from Xilinx Archives licensed under CC BY-SA 4.0

Image 62- Rich Mora

Image from the John East Archives

Episode 20 Part 2

Image 63- The Mars Rovers

Image from The Goldstar Gazette CC BY-SA 4.0

Image 64- Peter Teets

Image from unknown photographer CC BY-SA 4.0

Image 65-Esmat Hamdy

Image approved by Esmat Hamdy, licensed under CC BY-SA 4.0

Epilogue

Image 66- Dan McCranie

Image by permission of Dan McCranie Licensed under CC BY-SA 4.0

Cover

Cover by Al Perry

Front Cover Image of John East from the John East Archives

Back Cover from top left to right: Jayshree Ullal- see Image 30, Jen-Hsun "Jensen" Huang- see image 31, Robert Noyce- see image 12, Steve Jobs- see image 38, William Shockley- see image 9, Lisa Su- see image 32,

Inside Front Flap, John Kennedy Image from unknown photographer, NASA CC BY-SA 4.0

Index

INDEX

OK

Bell, 11, 27–28, 35, 45, 202, 271–273, 303
Bipolar, 57, 95–96, 99, 107, 122–125, 302
Birkner, John, vii, 219, 232, 312
Blanchette, Gene, 16, 116
Blank, Julius, 24, 121, 131–132
Brattain, Walter, iii, 28, 35, 308
Briggs, Jerry, 2–4
Brocade, 160, 17
Brooks, Don, 119

C

Carey, John, iii, 13, 15, 21–22, 122, 144, 307
Chang, Morris, vi, 209, 213, 216, 245, 249, 311
chip, x, xiii–xiv, 14, 42–43, 57–58, 67, 76, 207–208, 230, 232–233, 237, 255, 295, 304
Chua, HT, vii, 219, 232–233, 312
Cisco, 124, 147–148
Class, 10, 13–14, 77, 240, 272
Clos de la Tech, 146

CMOS, 103, 123, 145–146
Conner, Gene, 135, 151
Core memory, 56–58
Corrigan, Wilf, v, 109–110, 112, 115, 230, 309
Cotter, Ruth, 140
Crippen, Dick, 130
Cygnet, 79

D

Data General, 246, 249, 316
Davidow, Bill, vi–vii, 156, 159, 169, 222, 235, 248, 257, 265, 310, 312–313
Design win, 71, 160, 163–165, 168–169
DOS, 176–177
DRAM, 67–68, 71, 75–76, 156, 205, 245
DSP, 219, 250–251
DTL, 96–97

E

EEPROM, 80, 316
EPROM, 68, 76, 80, 205,

F

G

H

I

J

K

L

M

N

O

P

T

U

V

W

X

Z

CPSIA information can be obtained
at www.ICGtesting.com
Printed in the USA
BVHW092359100422
633743BV00005B/22/J